Dictator – How to Unfuck Your Leadership

Matthew Jacques

DICTATOR: How to Unfuck Your Leadership—*A System for Managers Carrying Responsibility Without Authority*

Published by: Matthew Jacques, Darwin NT 0800

ISBN Paperback: 978-1-7644795-0-9

ISBN E-book: 978-1-7644795-1-6

ISBN Hardback: 978-1-7644795-2-3.

Disclaimer

This book is based on the author's personal experiences, observations, and professional judgement. It is a record of lived experience. Everything in these pages reflects what I saw, heard, felt, and understood at the time. It is written from my perspective alone, shaped by memory, context, and the impact those experiences had on me as a person and as a leader. The events, conversations, and situations described in this book are not presented as findings of fact, legal conclusions, or objective assessments of the intent, character, or conduct of any individual or organisation. They are my recollections and my interpretations of how those moments were experienced and processed by me, whether accurately recalled or imperfectly perceived. References to leadership behaviour, workplace culture, and organisational systems are made only in the context of their effect on me and others as I experienced it. They are not claims of universal truth, nor allegations of wrongdoing. Where individuals or institutions may be recognisable, descriptions are limited to what I personally observed and the impact those experiences had on my wellbeing, judgement, and approach to leadership. This book is not an investigation, a complaint, or a verdict. It is a personal account of leadership under pressure, the consequences of systems, and the cost of decisions lived from the inside. It does not attempt to present the full perspective of all parties involved nor does it seek to assign moral, legal, or organisational responsibility. All names used throughout this book, with the exception of my own, are pseudonyms. Identifying details, roles, timelines, and contextual elements have been altered where necessary to protect privacy and confidentiality. Some individuals consented to being portrayed in this work; however, names and identifying details have still been changed to ensure confidentiality and reduce the risk of unintended harm. This book should be read as a narrative of experience and reflection, not as a statement of fact about the intentions or actions of others beyond how they were perceived and processed by the author. This book does not provide legal, medical, psychological, or professional advice. The author and publisher disclaim any liability arising directly or indirectly from the use or application of the material contained herein. Responsibility for decisions and outcomes remains with the reader.

Contents

Who This Book Is For – and Who It's Not

This book is written for people operating inside real systems, not ideal ones. It is for middle managers working inside large organisations, military, government, corporate, or enterprise, who carry responsibility without authority, pressure without protection, and outcomes without real control. It is for the people expected to deliver, lead, and absorb consequence while operating inside structures they did not design, do not own, and cannot easily change. These people sit between strategy and reality, translating decisions made elsewhere into results they will be held accountable for, shielding the people below them from fallout while quietly managing the damage created above.

It is also for CEOs, founders, and entrepreneurs who have outgrown personality-led management and brute-force effort. Leaders who have learned that constant availability does not scale, that hustle eventually breaks teams, and that what worked when the organisation was small fails once complexity arrives. Whether you operate in the middle of someone else's system or at the top of your own, the pressure lands the same way. Outcomes depend on you, but effort alone no longer carries the load. What is missing is not intent or drive, but a structure that can hold under pressure.

I built the Operator System under real conditions, balancing the demands of a growing business alongside service in the Royal Australian Navy. It was forged in environments where failure carried consequence, time was finite, authority was constrained, and leadership mistakes were felt immediately rather than explained later. This system exists because leadership had to work when resources were limited and personal effort could no longer compensate for structural weakness.

This book does not exist to energise or reassure. It does not trade in charisma, influence tricks, or personal branding. The Operator System is a leadership and management framework designed to build structure so you, and the people around you, can operate effectively without relying on heroics, constant intervention, or burning yourself into the ground. Structure holds when pressure arrives.

This book is for you if:

- You are accountable for outcomes but do not control the system above you.

- You are exhausted from compensating for organisational chaos with personal effort.

- You want clarity, standards, and rhythm instead of constant firefighting.

- You believe leadership should reduce pressure rather than push it downward.

- You are willing to sit in discomfort to build something sustainable.

This book is not for you if:

- You are looking for validation instead of responsibility.

- You want leadership to feel good more than you want it to work.

- You believe culture fixes itself through slogans and positivity.

- You want a system that removes moral weight and hard decisions.

- You expect clean answers, tidy endings, or guaranteed outcomes.

This book is written from inside the military, but it is not a military book. It does not exist to prosecute grievances or settle scores, even though it does not look away from pain, failure, or consequence. What you are reading comes from lived experience, but it is not centred on me. It is centred on the system that emerged when pressure removed the option of shortcuts and leadership had to function without excuses.

The Operator System did not appear fully formed. It was built over time, shaped by hard lessons, failed assumptions, and exposure to a wide range of leadership and management thinking. Ideas from people such as Dan Martell, Jack Delosa, Alex Hormozi, and others influenced its development, but this is not a stitched-together collection of borrowed concepts. What remains here is the version that survived contact with real environments, after theory met constraint, authority met limitation, and leadership stopped being abstract.

What you are reading is a load-bearing system. It is designed to function under pressure, in the absence of perfect leaders, and in conditions that are hostile to comfort or certainty. It is not elegant for appearance, and it does not promise reassurance. It exists because it works. If you are prepared to engage with it on those terms, it will give you something solid to stand on. If not, it may unsettle assumptions you have learned to protect. Either way, it does not soften from here.

Acknowledgements

Kim and Rhiannon, you carried far more than your share. While I was deployed, away, or consumed by a career that rarely respected balance, you did the work that never appears on a resume. You raised our children. You held the home together. You absorbed the absence, the uncertainty, and the quiet weight that comes with loving someone who is often not there. That labour is invisible to most people. It has never been invisible to me. It matters, and I am deeply grateful for it.

Addison, Rubi, Ripley, and Austin, you grew up inside a life that demanded resilience before you had any choice in the matter. You missed moments with me that you should have had. You lived with the downstream cost of a system that took time, energy, and attention away from home. I hope this book helps you understand not just what I did, but why I wrestled with it, and why I continue to question it. Everything I know about responsibility, consequence, and leadership is inseparable from being your dad. I love you all more than I will ever be able to express.

To my broader family, and especially my mother, much of what you read here will be new. This book contains experiences I did not speak about at the time, not because they were insignificant, but because carrying them quietly felt easier than explaining them. Reading this may change how you understand parts of my life, my choices, or my silence.

If parts of this are confronting, that is not because the truth is exaggerated, but because it was never shared. This book is not written to assign blame. It exists to be honest about what sustained pressure does to people when it goes unexamined for too long, and why I became who I became in the process.

To anyone reading this who is struggling with their mental health, particularly inside Defence or similarly rigid systems, I will say this plainly. Do not carry it alone. Silence is not professionalism, and endurance is not strength when it is slowly hollowing you out. Ask for help early. Speak while you still can. Taking responsibility sometimes means admitting you are not okay.

This book is written with care, but without softening what happened or what it cost.

I love you all.

Prologue

The $226 Fine

You don't get punished for being wrong. You get punished for being inconvenient. — Unknown

In March 2025, I stood in a mock courtroom inside the Walker Room of the Command Building at HMAS *Coonawarra*. Across the table sat the Commanding Officer and Senior Naval Officer Northern Australia. Between us lay a single sheet of paper.

A $226 fine.

That was the formal record of what the system could see.

On paper, the offence was simple. I had called my superior officer a dictator. Twice. In front of witnesses. I had slammed a door hard enough to rattle the building. The behaviour was logged, processed, and priced accordingly.

What the paper did not capture was the fourteen months that led there.

Nine months earlier, fines were irrelevant. I was vomiting every morning before work. Not occasionally. Not when things were bad. Every fucking morning. My body had started rejecting the environment long before my head caught up.

I was planning which corner in Berrimah I would drive my Mustang into. I knew the road. I knew the speed. I knew the barrier. I had rehearsed it enough times that it stopped feeling dramatic and started feeling procedural.

I had been diagnosed with an acute mental health disorder. I could not see a future that didn't involve either destroying everything I had built, or destroying myself.

By the time I stood in that room, the labels attached to my name looked very different.

Chief Petty Officer.

Dux of my leadership course.

Recipient of two national safety awards.

Leader of a team who described their workplace as the best they had ever experienced.

Those facts sat neatly together on paper. They made sense to the system.

What didn't make sense was how close those outcomes sat to the edge.

I looked at the $226 fine and understood something clearly for the first time in a long while.

That fine was not punishment. It was a receipt.

It was the price the system charged when someone stopped absorbing damage quietly and became inconvenient instead. It marked the moment when silence finally broke, not because of courage or confidence, but because there was nothing left to give.

This book is not about that moment.

That moment was just the symptom.

What this book traces is the failure that came before it, and the structure that had to be built in response.

I did not recover by standing up to anyone. I did not think my way out of that hole. I did not heal through insight, optimism, or reframing.

Every one of those options had already failed.

I survived because I built a system that did not rely on goodwill, personality, permission, or functional leadership. I built it because the organisation I was operating inside had removed every other reliable option.

Pressure was constant. Authority was constrained. Consequence landed downward. Silence was rewarded until it became dangerous. When leadership failed, the system did not correct itself. It closed ranks.

So I stopped expecting it to.

The Operator System was not designed to make me a better leader. It was not built to inspire people or improve culture. It exists because operating inside a broken system without structure will eventually kill you, either slowly or instantly.

This book follows the path from breakdown to that courtroom, and then beyond. I went from operating under leadership that hollowed people out to building a way of working that could hold under pressure without relying on heroics or sacrifice.

If you are reading this while barely holding yourself together, managing chaos with personal effort, or planning your exit from a system that is grinding you down, you are not reading a recovery story.

You are reading a survival manual.

If you carry responsibility without authority, outcomes without control, and consequence without protection, this will be familiar in ways you may not enjoy.

It gets dark. But there is a way through.

Nothing from here gets fucking softer.

One

The Slow Death

You only realise how bad leadership is once you've experienced good leadership.   — Unknown

To understand how I ended up standing in a mock courtroom holding a $226 fine, you need to understand the descent that led me to that moment. Not the moments that look dramatic from the outside, but the slow erosion you explain away while it's happening, until you're suddenly facing the Commanding Officer and realising you no longer recognise your own life.

Before I tell you about the most toxic leader I ever served under, you need to understand the leader who shaped me.

Lieutenant Commander Mike Brigg; Officer in Charge of Fleet Logistics Support Element Darwin from January 2021 to January 2023. One of the best operators and leaders I ever served under.

People above him called him the businessman or the policy man. From where I sat, he was the most people-centred leader I'd seen. He knew his people properly, from the newest Able Seaman (AB) still finding their feet to the sailor carrying a load heavy enough to break most people. He understood their families, their pressures, their limits, and what they needed to deliver without burning themselves into the ground.

He led without losing his humanity and managed without forgetting that people were not machinery. Decisions were made with the mission in mind, but never at the expense of the people doing the work. Standards stayed high, pressure was real, and accountability was clear, but it was never enforced by fear. He proved, day after day, that you did not have to trade people for performance. You could deliver both, and you could do it fucking well.

When he posted out in January 2023, we did not just lose a boss. We lost the spine of our culture.

What most people never saw was how much of that leadership I was exposed to. My desk sat outside his office, close enough to hear the shift in his voice when senior ranks pushed their latest 'COGI' down the line. Commanding Officer Great Ideas were usually dreamed up by a bunch of fucking people far removed from the coalface, and never accountable for the fallout.

I saw Brigg absorb those hits so the rest of us didn't have to. He stayed calm, measured, and unflinching. He held ground without raising his voice, and every time he did it, I took note. Those moments taught me more about leadership than any course I ever took.

For two years he mentored me without ever calling it mentoring. I learned by working alongside him and paying attention to how he held himself under pressure. The structure, the presence, the backbone, and the humanity were all there, carried naturally rather than performed.

In that whole time, we only had one moment that went sideways, and it was because of me. I said yes to supporting the Chief of Navy. At the time, it felt automatic. When someone with that much brass asks for something, senior sailors say yes and sort it out later. I hadn't learned yet that saying yes to the wrong thing creates more damage than saying no early.

Brigg pulled me into his office. There was no yelling and no theatrics, just a level of disappointment you could feel settle into your bones. He walked me through exactly why I'd made the wrong call. Our job was to support Patrol Boats, not whatever swing-dig the Chief of Navy wanted to run. There were no business rules for it, and my yes had kicked off a logistical circus.

Stores were dragged out of Darwin and handed to what we openly call the civilian branch of Defence, the Air Force. They organised with the urgency of the Northern Territory, more affectionately known as Not Today, Not Tomorrow, Not Tuesday, and certainly not Thursday. By the time the gear finally crawled toward East Arnhem so the Chief of Navy and his entourage could have their welcome-to-the-sea moment, the damage was already done. Finance had no idea who was paying, movements seized up, and the system bent out of shape because I hadn't stopped long enough to think through the second- and third-order effects of saying yes to the wrong bloke.

Brigg didn't care who he had to push back on. Rank wasn't the point. Getting it right was. The mission mattered, and the people doing the work mattered even more. That conversation changed how I understood backbone. Being on the receiving end stung, but it also locked in my respect for him.

What gave Brigg that presence was fifteen years in senior corporate roles before returning to Navy. His last role was Head of Logistics for NBN, carrying responsibility and pressure at a scale most people never see. You could hear that background in the way he thought problems through, planned work, and communicated intent. His thinking wasn't Navy-thin. It was commercial, disciplined, and built to scale. I paid attention because I was already carrying responsibility outside the uniform, and I knew those lessons would matter, and they did.

So when he walked out the door for the last time, something left with him. The place didn't fall apart overnight, but it sure as shit felt a little emptier.

And then she arrived.

Lieutenant Commander Sarah fucking Carrigan.

At first glance, nothing looked wrong. Her superiors called her a people person. I told myself to be optimistic. For the first month or two, it felt workable. Not Brigg-level. Just enough to convince you it might last.

Then the rot set in.

It happened slowly, the way timber softens from the inside before anything collapses. At first, there were small changes you could explain away. A shift in tone. Less engagement. Decisions that felt off but not yet dangerous. The culture that had been solid under Brigg started to weaken, not enough to point at directly, but enough that you felt it day to day.

The early signs were noticeable but easy to excuse. Mood swings. Tears in the workplace. Personal issues brought straight into the office. Conversations repeated because she'd forgotten having them. Then there was the disappearing act.

She'd arrive around 0800 as if everything was normal, drop her bag, and make a point of how busy she was. By 1000 she'd be gone. No meeting. No tasking. No explanation.

Some days she'd reappear mid-afternoon, drift through the office asking questions, then disappear again. We were left trying to run a unit with an Officer in Charge who treated the workplace as somewhere she passed through rather than led.

On a good day, she was in the building for three and a half hours, and even that was generous. Much of the time she was present, she was spiralling about issues that had nothing to do with the work. The rest of us were left running the unit under an Officer in Charge who treated the office like a bus stop she might or might not pass through, depending on her mood.

What made it worse was that the authority never left with her. She was gone most of the day, unreachable, and then she would reappear questioning decisions we had no choice but to make in her absence. We were expected to keep the unit running while also accounting for choices made without guidance and after the fact.

I gave her more grace than I should have. She was a single mum with a life that looked permanently stretched thin. Anyone could see the weight she was carrying. I understood it because I knew what it felt like to have home bleed into work and to try to carry both without dropping something important. I kept telling myself it would settle, that she'd find her feet, and that patience would be enough.

Over time, it became clear we were propping up someone who didn't have the capacity to carry the role she'd stepped into. That realisation didn't come all at once, and I resisted it longer than I should have. I kept telling myself patience would change the situation. Instead, the gaps widened, the confusion deepened, and the damage started to compound.

The pattern didn't correct itself. It hardened. Every Monday morning I'd open my inbox to find another forwarded email waiting for me. No subject change. No note. No direction. Just a safety task meant for her role quietly dropped into mine.

At first, I checked in. I'd ask, "Ma'am, do you want me to action this?" Sometimes she'd nod. Other times there was nothing at all. By Friday, I'd be called into her office and asked why the task hadn't been done. I'd point out that it had been directed to her and that I hadn't been given guidance, but it never mattered. The expectation was that I should have known what she hadn't said. There were no deadlines, no direction, and no tolerance for the gap she'd created. Just blame.

After a while, I changed how I operated. It became easier to absorb the impact than to keep pushing back through her fog. Direction was absent, expectation was constant, and the consequences landed wherever there was room.

Then came September 2023, the moment the floor shifted.

By then, every small issue triggered the same response. Late paperwork, minor admin errors, or reasonable questions were met with immediate emails sent without pause or restraint. There was no reflection and no attempt to separate personal frustration from professional responsibility. Responsibility was always pushed outward.

But even with all that, nothing prepared me for what came next.

I was on parental leave in Cairns with my newborn, standing in the kitchen with him in my arms, half-asleep and smelling of nappies and formula. My partner, Rhiannon, was in the other room. It should have been quiet.

I opened my email. A dumb fucking mistake.

What followed was an accusation delivered without warning, context, or care. I was accused of creating a boys' club, of negligence, of failing to resolve tension in my department, and of discriminatory behaviour. There was no conversation attached to it, no attempt to clarify what had supposedly happened, and no regard for where I was or what I was doing when it was sent.

The issue itself was ordinary. A disagreement between two sailors who had carried baggage from a previous posting into ours. There was no favouritism, no misconduct, nothing that warranted the language she used. None of that mattered. She had already decided on the story, and once she committed to it, that version became fixed.

Reading it, I felt the shift straight away. The moment you recognise that someone with power has decided you are the problem, and no longer needs evidence to make it real.

The message underneath was clear. My leave didn't matter. My family didn't matter. I had failed, even while I wasn't there.

I spent two days pulling myself apart. I should have been present with my newborn. Instead, I paced the house, replaying her accusations on a loop I couldn't shut down. Rhiannon saw it immediately. She stepped in, grounded me, and carried the weight I couldn't get a handle on in that moment. Without making it bigger than it was, she kept things steady while I spiralled. Even so, the damage was already done. That email lodged itself deep and didn't move.

When I returned to work, nothing was said. It was as if it had never happened. Cold stares. Business as usual. Like the grenade she'd lobbed into my life was just another day at the office.

That was the point where something shifted internally. I realised the trust I'd been operating on was gone, and I hadn't noticed it draining away.

From then on, I understood one thing clearly. I was no longer safe in a role I had once thrived in under Lieutenant Commander Brigg.

Two

The Vision, Mission, and Values that Didn't Matter

The appearance of alignment is often mistaken for alignment itself.
(paraphrased)   — Patrick Lencioni

Around that time, the mental cost started to register. The constant interruptions, shifting expectations, and absence of clear direction had been grinding away for months. It was taking more effort than it should have to keep things moving. Among the bullshit, there was still a part of me that wanted the work to function properly. I still believed leadership could be done well, that culture could be repaired, and that clarity, applied simply and consistently, could steady a team.

When Vision, Mission and Values are done properly, how people operate changes. It removes ambiguity and gives a unit a shared reference point for decisions, standards, and behaviour. It's not corporate wallpaper. It's the underlying logic of how a team functions day to day.

By that point, I could see where Fleet Logistics Support Element Darwin (FLSE-D) was heading. The cracks were already showing, and without something concrete to anchor the unit, they were only going to widen. This work felt like a chance to reset direction before the damage became irreversible. Not because words fix problems on their own, but because clear standards give people something real to work from. That thinking would later become the foundation of what I now call the Operator

System, but at the time it was simply the most practical way I knew to stop a team from drifting.

After months of absorbing someone else's disorder, the Vision, Mission, and Values work gave me something concrete to focus on. It was practical and bounded, and gave me a way to reinforce the standards Brigg had established before they diluted any further.

The work pulled me back into the job itself. I was spending less time reacting and more time putting structure in place.

I brought the whole unit together. Every rank and every department. Storbies (warehouse attendants for boats and ships), chefs, junior sailors, senior leaders. We ran ten hours of group sessions, followed by weeks of drafting and coordination to make sure the result reflected who we actually were, not who someone else thought we should be.

People engaged with it.

The tone in the room shifted. Conversations that had been quiet for months started happening in the open. Sailors spoke up about issues they'd been carrying without a place to put them. Ideas surfaced that hadn't previously made it past the mess table. The work created space for people to contribute without being shut down or second-guessed.

What we built reflected that. A vision centred on people doing meaningful work. A mission anchored in proactive, human-centred support for operational outcomes. Values grounded in trust, respect, openness, and harmony, written in language the unit actually recognised as its own.

Looking at them now, I hate them. They're the kind of one-word, 1970s corporate values that look tidy on a wall and say fuck all about how people are actually expected to behave. I don't believe in that style of work anymore. But at the time, they mattered because the people who built them mattered. Their fingerprints were on every line, and that ownership counted for more than polish.

When November rolled around, we presented the draft to her.

She skimmed it without looking up. There were no questions and no engagement with what was in front of her. She closed the document with her hand and told me to change it.

I asked what needed to change.

She said the vision.

That was it. No explanation of what wasn't working and no indication of what she wanted instead. The work went back into motion without a target, and the ambiguity landed on the same people who had already invested the time to build it. Momentum stalled. Frustration crept in. The energy in the room shifted as effort was redirected without purpose.

So we revised it and sent it back.

She dismissed it without explanation. We revised it again, trying to interpret feedback that hadn't actually been given. This time her response contradicted what she'd said the round before, sending us back over ground we'd already covered. We took another pass, only to be pushed toward the very ideas she'd rejected at the start, as if the previous weeks of work had never existed.

Eventually, the pattern became clear. The revisions weren't about improving the work. They were about removing anything that hadn't originated with her. Each round stripped away more of what the team had built and replaced it with language that carried her voice instead.

What we had put together stopped feeling like it belonged to the unit. As decisions moved behind closed doors, ownership thinned out. People who had contributed early began to step back, not out of apathy, but because there was nothing left that asked anything of them.

By February 2024, I could see the limit of what the process was going to allow. I wasn't finished, but I was no longer willing to keep pushing energy into something that was being hollowed out faster than it could be built.

I sent the email.

Dear Ma'am,

I hope this email finds you well. I am writing to inform you of my decision to step back from further discussions and involvement regarding the FLSE-D Mission Statement.

Over the course of our work, I have dedicated a significant amount of time and effort towards contributing to this project. However, after much consideration, I've come to realise that for my own personal wellbeing, it is necessary for me to remove myself from this process at this stage.

Please understand that this decision was not made lightly. I have great respect for the work being done and the vision that the team is striving to achieve. I believe that the mission statement is a crucial component of our collective goals, and I've been honoured to contribute to its development thus far.

While I am stepping back from the discussions, please know that I remain supportive of the project's objectives and am keen to support the finalised mission statement. Once it is finalised, I am more than happy to buy-in and align my efforts with the established direction.

I wish the team the very best in finalising the mission statement. I am confident that with the collective expertise and dedication, the outcome will be both impactful and reflective of our shared values.

Thank you for your understanding and support in this matter. Should you need to discuss this further, please feel free to reach out to me.

Warm regards, PO Matthew Jacques

Read between the lines, and it was obvious what I couldn't say outright.

"Personal wellbeing" was the safest language I had for being overwhelmed by a situation I no longer had any control over. "Respect for the work" was cover, not truth. And "buy-in" wasn't agreement; it was withdrawal. It was the point where I stopped trying to shape the outcome and focused instead on limiting the damage to myself.

What situations like this force on you is restraint. You end up writing carefully when everything in you wants to be direct. You send polite, professional emails while carrying thoughts you know you can't put on the page.

What I wanted to write was simple. You have no idea what you're fucking doing, and I won't be part of this shit show anymore.

That wasn't an option. Not in the military, and not to the person who still controlled my posting, my reports, and my future. So the language stayed measured, the tone stayed respectful, and the truth stayed unspoken.

People started reaching out, asking why I wasn't leading the work anymore, and there was no explanation to give them. The project continued under her direction, but engagement faded. Attendance thinned, contributions dropped off, and the work shifted from something people cared about to something they simply complied with.

From February through June, the work effectively disappeared from view. There were no updates shared, no drafts circulated, and no indication of where it was heading. Access narrowed. The people who had built the original version were no longer involved, and the work moved behind closed doors. Whatever direction it took from there, it was done without consultation and without the unit that had been asked to contribute in the first place.

Just control.

She eventually called the unit together for the unveiling. We gathered downstairs in the office, crowded around the big TV, waiting to see what had come out the other side of months of work we were no longer part of.

She stood at the front like she was about to announce her coronation.

When she opened with the Vision, Mission, and Values, there was no acknowledgement of the team who had built the original version. No reference to the workshops. No recognition of the time or effort that had gone into it. She presented it as if it had been created in isolation, detached from the people who had contributed the most.

She moved on to the new vision and read it out as if the words themselves were self-evident.

"Support Home and Away Ported Patrol Boats with an Unreasonable Level of Support."

The word landed badly, at least it did for me. Unreasonable. Not as a challenge or an aspiration, but as an instruction. A way of signalling that people were there to be spent. It wasn't grounded in the reality of the work or the limits of the team. It read like permission to take more without understanding the cost, written from far enough away from the coalface that the consequences would land on someone else.

She took a people-first vision built on trust and recast it as a demand for sacrifice. What had been intended to protect sailors was reframed as an expectation that they would give more, regardless of cost.

The room stayed flat. No side comments. No visible reaction. It wasn't respect. It was acceptance. The kind that comes when people already know their input no longer carries weight.

After the brief, PO Gunn leaned over and kept her voice low. "Well, that was a waste of our fucking time."

She was right. And it wasn't just about the hours lost. The message was clear enough without being said.

The waste wasn't the brief. It was what followed. In one sentence, months of shared effort were stripped of ownership. The work didn't belong to the people who built it anymore. It belonged to rank. The language made that clear. From that moment on, we weren't shaping how support was delivered; we were just executing someone else's words. Our role shifted from contributors to capacity. Labour without voice. The job still existed, but the meaning had been hollowed out of it.

The work was no longer ours, and neither was the voice that came with it.

Three

The Collapse in Plain Sight

The greatest danger in times of turbulence is not the turbulence itself, but relying on outdated thinking to navigate it.   — Peter Drucker

FLSE-D wasn't just under strain; it was tightening in on itself. You could feel it when you walked in. Conversations became shorter. People were more careful about what they said and who they said it around. The unit still functioned, but it no longer felt settled.

Over time, the same patterns kept repeating. Early assumptions that things would smooth out stopped making sense as nothing actually changed. The instability wasn't tapering off or correcting itself. It became clear this was simply how she ran the place.

Decisions stalled because she couldn't commit to them. Priorities shifted based on her mood rather than the work in front of us. Minor issues could absorb hours of attention, while genuinely important matters drifted unattended because she didn't understand them and made no effort to. Junior sailors learned quickly that asking questions brought scrutiny rather than guidance. Senior sailors started taking work home because waiting for direction inside the office meant nothing would move.

Workloads increased as clarity fell away. Tasks kept moving, but direction didn't. People still showed up and did the work, but there was no sense that anything above us was actively guiding it.

What made it harder was knowing the system around her absorbed the impact. Rank buffered her from consequence. Process diluted accountability. Over time, people learned that raising concerns only created more exposure for themselves, so they stopped doing it. Heads went down, risk went quiet, and the unit adjusted around the dysfunction instead of confronting it.

At some point, it became clear that my relationship with the organisation had changed. Not in a dramatic way, and not all at once, but enough that I couldn't ignore it. The Navy I'd built my adult life around no longer felt like the place I was operating in day to day.

I'd always assumed I'd leave on my own terms, at a time that made sense, with a clear sense of pride in what I'd given and what I'd taken from the job. Instead, I reached a point where staying was starting to cost more than leaving ever would.

In March 2024, I submitted my discharge.

It wasn't a protest and it wasn't a bargaining move. It was a practical decision made to preserve what I still had left. For twenty years, the Navy had been more than a job. It had shaped how I saw myself and how I operated.

By that point, the way I believed leadership should be exercised no longer aligned with the environment I was working in. The standards I was prepared to accept, enforce, and live by had diverged from what was being normalised around me. Staying meant compromising things I wasn't willing to compromise, and I wasn't prepared to do that quietly.

Leaving wasn't as simple as signing a form. I was locked into a Return of Service Obligation until August 2025, which meant that even after deciding to walk away, I still had to remain in that environment for another seventeen months. It was time I had to endure under leadership I no longer trusted, in an organisation that I no longer aligned with, while trying to protect whatever parts of myself hadn't already been ground down.

Accepting that reality forced a practical conclusion. If I was going to last that long, I needed work she couldn't touch. Something outside her control that produced real outcomes. I wasn't prepared to keep absorbing the same dysfunction without putting something solid into the world alongside it.

So I built the FLSE-D Safety Magazine.

Not because I was directed to, and not to impress anyone. I built it because it was bounded work with a clear purpose. It dealt with real issues the team faced and could be produced without being pulled apart or redirected. It gave the unit something practical to rely on while everything else remained unsettled.

Safety, if we're honest, is usually about as exciting as watching concrete dry. Most safety messages read like you're being scolded for trying to stay alive. I wanted to change that. I wanted something people might actually open without sighing or rolling their eyes. So I made it simple and human. Short pieces people could skim. Real examples sailors would recognise. A tone that sounded like one of us, not a committee.

I put the whole thing together myself. Design, writing, editing, the lot. Most of it happened late at night or on weekends when the house was quiet and I had enough headspace to actually think. And strangely enough, instead of draining me, it lifted me. It gave me a reason to engage again. Something that wasn't tangled in chaos.

The surprising part was how quickly it caught on. People actually read it. People used it. Supervisors started including it in their briefs. Sailors brought it up in conversation. It worked because it wasn't corporate fluff. It was just a bloke trying to look after the people he worked alongside.

Which, of course, meant she disliked it immediately. There was no concern about risk, compliance, or quality. The issue was simpler than that. It hadn't come from her. Anything that worked without her fingerprints on it was treated as a threat rather than an achievement.

She started by trying to shut it down quietly. Comments about "reviewing communication channels", questions delivered in that loaded tone where the expected response was to pull back. When that didn't work, the pressure increased. The intent didn't need to be stated. It was obvious in the way the scrutiny tightened and the space around the work narrowed.

So I shifted it.

I renamed it the HMAS *Coonawarra* Safety Magazine, widened the audience, and moved it outside her control. Once it sat beyond the FLSE-D boundary, it was no longer hers to interfere with. It became visible to other departments, and that visibility changed the calculus. She wasn't going to challenge work she didn't understand across an audience she didn't command.

Once the name changed, she backed off and acted as if it wasn't happening. I kept going. Edition after edition, the work continued and expanded beyond the original scope. Over 2024 and 2025, I produced twenty issues, each one refining what came before.

The content stayed simple and practical. Real examples. Clear guidance written in plain language. No ego and no noise. Just material that made people's working lives a little safer and a little easier.

That work would later be formally recognised.

Across 2024 and 2025, I was awarded the Chief of Navy Safety Award, the CMDR Dave Allen Shield, and the COMCARE Award for Outstanding Contribution by an Individual. The citations focused on the magazine itself, the increase in engagement during safety activities, and measurable improvements in reporting culture and operational outcomes at Coonawarra.

It was formal recognition of work that had never been supported internally, delivered without endorsement and sustained entirely through consistency.

She never wanted it to exist. It existed anyway.

What mattered wasn't the recognition itself. It was the confirmation that the work held up on its own terms, without permission or protection.

Nothing else changed. FLSE-D didn't suddenly become easier to operate in. Her behaviour continued, and the unit kept adjusting around it. The work gave me something solid to keep building, but the day-to-day reality of serving under her remained volatile and unpredictable.

Which brings me to ANZAC Day 2024.

ANZAC Day sits apart from the rest of the calendar. It carries a different weight, one that changes how people behave and what they notice.

25 April 2024. Darwin had that warm late-afternoon feel, the kind that settles into your shoulders without pulling you out of the moment. By about 1530, Monsoons on Mitchell Street was packed, the way it always is after the service. Sailors in white ceremonials, veterans wearing medals that carried weight most people never have to shoulder, families drifting between tables, the usual Darwin crowd moving between laughter, stories, and the usual shit talking that goes on.

Monsoons wasn't just a bar to me. Neptune NT ran security there, and I owned the company. I knew the venue, the staff, and how situations escalated when alcohol and emotion mixed. It was familiar ground, and I understood the dynamics of the place without needing to think about it.

I was talking with Lieutenant Commander Mick Cutlis, someone I'd served with on HMAS *Leeuwin*. He was now the Maritime Logistics Officer on HMAS *Adelaide*. We were talking at the usual catch-up when he mentioned something that had happened the day before. She'd changed *Adelaide*'s storing times without telling him, leaving him blindsided and scrambling to adjust.

It was frustrating, but familiar. The kind of disruption you learned to account for. Certainly not something either of us was interested in carrying into an ANZAC Day beer.

Then she walked in.

She was in civilian clothes, but the posture was familiar. Jaw set. Eyes scanning the room in that restless way she had, as if she was already keyed up before a conversation had even started. She spotted Cutlis almost immediately and moved straight toward us.

There was no pause to read the room. No adjustment for the setting. In a space filled with uniforms, veterans, families, and the quiet weight that comes with the day, she closed the distance with a focus that felt out of place and unnecessary.

Before I could even set my drink down, she was already confronting him. It wasn't a discussion and it wasn't measured. She unloaded directly, personal and unrestrained. She did it on ANZAC Day, in a crowded venue, in front of people who knew me and expected anyone connected with Navy to carry themselves with basic restraint.

I stepped in immediately, positioning myself between them the same way I had countless times in the early hours of the morning with drunk punters who'd lost control of their tempers. I guided her back slightly and kept my voice level, firm enough to register without feeding what was already building in her.

"Ma'am, this isn't the place. Think about the day. Think about who's around."

She looked at me like I'd insulted her. Sharp, cold, completely oblivious to how far over the line she'd already stepped. She tried to push past me to get back to Cutlis, so I settled my stance and let the security operator in me take over.

"Leave now, or security will take over. Your call."

She paused for a second, caught between pushing forward and the boundary in front of her. Whatever calculation was happening didn't last long. She turned away abruptly and left, cutting through the crowd and out the door.

The reaction was immediate and unfiltered. People swore. Heads shook. More than one sailor said exactly what they thought about her and what had just happened. It wasn't subtle and it wasn't polite, but it was consistent. Everyone in the room had read the moment the same way.

They didn't need context or backstory. They didn't need to know about the disappearing acts, the mood swings, or the chaos she carried into the workplace. What they'd seen was enough. A senior officer losing control in public, on a day that carries weight whether you wear the uniform or not.

I'd spent months giving her the benefit of the doubt, explaining things away and telling myself this behaviour was temporary or circumstantial. I didn't recognise at the time how much self-doubt had already crept in, or how much effort I was burning trying to normalise something that wasn't normal.

Watching her lose control in front of a room full of people on that day stripped those explanations away. There was no framing left that made it acceptable, and no version of events that softened what had just happened. It wasn't pressure, fatigue, or a difficult posting showing itself. It was simply how she operated.

From that point on, I stopped pretending otherwise.

Four

When High Performers Become Targets

Strong leaders engage their critics and make themselves stronger. Weak leaders silence their critics and make themselves weaker.   — Unknown

By May, FLSE-D hadn't collapsed, but it sure as hell wasn't functioning properly. Work still moved and tasks still got done, but the bullshit never stopped circulating. People talked openly about her incompetence and the frustration it created, far more openly than they ever would have under someone who knew what they were doing. It wasn't whispered or careful. It was blunt, because there was nothing left to protect.

Senior sailors stopped wasting energy trying to fix her and focused on keeping their people intact. Most of their effort went into absorbing the bullshit coming down from above so their teams could still get work done. Junior sailors weren't subtle about it. They were already looking for postings out, counting days, talking openly about anywhere else being better than here.

FLSE-D's reputation didn't stay local. Word travelled to other bases, and everyone knew what was happening. Command did nothing. To be fair, she knew exactly how to perform upward. Briefs were clean. Optics were polished. From the outside, she looked exceptional. Inside the unit, where the work actually happened, it was obvious she had no fucking idea what she was doing.

The gap between how things looked and how they actually were didn't appear on paper. It showed up in the sailors everyone relied on to keep the place functioning.

I'd been carrying pressure for a long time. Navy demands a lot, and that wasn't new. On top of that, I was member separated, with Rhiannon and the kids back in Cairns. My relationship with the OIC was so fractured that even asking to go home drew scrutiny until I stopped asking altogether. By this point, I'd been home for a total of three days.

In March, my body started pushing back. I spent two weeks in hospital with pancreatitis, followed by emergency surgery to remove my gallbladder. It wouldn't be the last surgery of that year. I recovered and went straight back to work. That's what I'd always done.

At the same time, I was still heavily involved in the day-to-day running of Neptune NT. We'd just brought in an Operations Manager, Zak Cole, a former Chief Naval Police Coxswain, to take some of that load off me.

Around that period, I also joined a business coaching group called the Entourage. At the time, it wasn't about escape or reinvention. It was simply another attempt to sharpen how I operated and keep learning at a level that matched the pressure I was carrying. I didn't realise then how important that decision would become, or how much it would help me navigate what followed through the rest of 2024 and into 2025.

But none of that was what stood out.

I was shocked by the changes in the people around me. Sailors who used to turn up with energy now arrived, did the work, and left. Conversations flattened. Laughter thinned out. People weren't coming in because they wanted to be there anymore. They were coming in because it felt safer than drawing attention to themselves.

That pressure didn't spread evenly. It went where it always goes. Onto the sailors who carry the load when leadership goes to custard. The ones who close gaps without being asked, keep standards intact, and absorb uncertainty so others don't have to. They don't announce it. They just take on more until the cost starts showing.

In FLSE-D, two senior sailors carried more responsibility than most, and they carried it without needing to be told. Not because it was recognised or rewarded, but because that's how the place stayed upright. When decisions stalled, they closed gaps. When direction blurred, they provided it. When pressure landed, it landed on them.

PO Jacinta Gunn and PO Joel Howard were among the strongest operators in the unit. Their work didn't create noise, but it created stability. People relied on them because they were consistent, because they knew their jobs, and because they could be trusted to handle weight without dropping it.

Which brings me to Jacinta.

PO Jacinta Gunn worked in Support Operations. Historically, the role had been treated as domestic support. Jacinta turned it into something else entirely.

She effectively ran the financial and logistical control for twelve patrol boats. Stock management, cash handling, reconciliation, compliance. Every dollar that moved through those boats passed through her system. Nothing drifted. Nothing disappeared. Errors didn't compound because they didn't get the chance to.

What she built wasn't flashy. It was disciplined, repeatable, and reliable. The kind of work that doesn't draw attention until it's gone. In a unit already struggling with instability, her output quietly kept a critical part of the operation upright.

She didn't just keep it moving. She kept it disciplined.

Her reliability came from understanding the system end to end and treating accuracy as non-negotiable. The work held on its own, which made it indispensable to the unit and increasingly difficult for the person nominally in charge of it.

She wasn't immune to the environment the OIC created. Some mornings I'd see Jacinta sitting a little too straight at her desk, her eyes checking the OIC's door before she spoke. It was subtle, but it was there. You could read it without a word being said.

Her output never slipped, but the cost showed up elsewhere. You could see it in her face and hear it in the conversations we had. She started working from home more often, counting down to time off in a way she never had before. The standard stayed high, but it took more out of her to maintain it.

Then there was PO Joel Howard.

Joel was one of the strongest stores operators I've served with. He ran his department tight. High standards, clean records, and zero tolerance for anything that didn't add up. If something looked off, it didn't get ignored or worked around. It got addressed.

When Joel raised an issue it was because he'd already checked the facts and followed the trail far enough to know it mattered. He trusted the system to do what it was meant to do once the truth was put in front of it.

In May, he raised the issue of a serious stock irregularity. The sort of issue that, under Brigg, would have been handled quickly and through proper channels.

Within weeks, Joel was gone. There was no warning, no transition, and no explanation offered to the unit. He was simply removed.

That kind of removal doesn't happen casually. Not to a PO, and not without process. There was no paperwork shared, no explanation offered, and no conversation held with the unit. It simply happened.

Seeing what was happening to Jacinta and what happened to Joel made the situation hard to ignore. These were capable sailors who carried responsibility without needing supervision. Under Brigg, that capability had been used. Under the current OIC, it became something to manage around. The difference was visible.

What I learned, watching sailors like Jacinta and Joel, is that high performers don't need managing the way weak ones do. They need space, clear standards, and the confidence that their competence won't be punished. When you let them operate, they lift everything around them without being asked.

Under her, the opposite played out. Capability created friction instead of momentum. Confidence was treated as defiance. Integrity became inconvenient. Initiative narrowed the space rather than expanding it.

Jacinta stayed. Her work never slipped, but the way she moved changed. After Joel left, everyone else adjusted their behaviour accordingly.

I wasn't outside that environment. I was operating inside it. Without noticing at first, I started adapting the same way others had. Choosing words more carefully. Letting things slide that I would've challenged before. Shaping myself to fit the space rather than pushing back against it.

Over time, that adaptation carried a cost. The confidence I'd relied on dulled. Decisions took more effort. The work felt heavier than it should have. I wasn't angry or dramatic about it. I was just worn down in ways that crept up quietly and stayed.

I still turned up every morning and did the work. The environment hadn't improved, but the responsibility hadn't gone away. People still relied on the job being done properly, even when leadership wasn't providing anything to work from.

Operating under her meant effort rarely translated into stability. You could keep things moving, but nothing stayed settled. Sailors began narrowing their engagement, doing only what was necessary to avoid attention. The unit didn't fail in a visible way. It thinned out quietly as people protected themselves.

What stayed with me wasn't just her behaviour. It was the effect of operating inside it for too long. The days narrowed. Everything took more effort than it should have. Sitting in the car before work became part of the routine, not because I didn't know what to do inside, but because it took time to switch myself back on.

Somewhere along the way, the job stopped feeling like service. It became something to manage through.

By June 2024, that strain was showing. Focus slipped. Rest didn't restore anything. Even though the unit was carrying its own weight, I felt isolated inside the impact of it. I still turned up and did the work, because that's what you do, but it wasn't sustainable.

And there's only so long a unit can operate like that before someone outside the immediate chain notices.

In late June 2024, someone finally did.

That's where the Director of Navy Culture enters the story.

Five

When Navy Culture Shows up

The greatest enemy of knowledge is not ignorance; it is the illusion of knowledge. — Daniel J. Boorstin

The Director of Navy Culture didn't turn up because of us. We weren't special. We were just one unit on a longer list.

Navy Culture is an internal Defence organisation tasked with assessing and improving workplace culture across the fleet. It sits somewhere between audit and intervention and is brought in when leadership, trust, or psychological safety are in question.

They were running a 360° cultural survey across multiple units, and FLSE-D happened to be next. The assessment measured twelve areas of workplace culture: values, communication, trust, psychological safety, accountability, teamwork, recognition, and whether people actually enjoyed coming to work. Not abstract theory. The basic conditions that determine whether a team functions or quietly rots from the inside.

The scoring was simple.

Fifty percent meant a category was effective. Anything below that meant it wasn't.

FLSE-D didn't score above fifty percent in a single category.

On paper, we weren't a functioning unit at all. Just a cultural clusterfuck with a nameplate.

The culture hadn't just dipped. It had already collapsed. The survey didn't reveal anything new; it simply put numbers against what every sailor in that building had been living for months. It quantified the tension you could feel the moment you walked into the workspace.

When the report started circulating through the management team, I didn't need to read it twice. I wasn't surprised. We finally had confirmation of what had been impossible to say out loud without consequences. You'd have to be blind fucking Freddy or brand new to miss the strain. Anyone who'd spent more than five minutes in that office could feel it.

Data doesn't soften the truth to spare feelings or rank. It doesn't negotiate with ego. It lays everything out plainly, and what remained was clear evidence of how far we'd fallen.

Two weeks later, on 20 June, we were marched into a full-day cultural session run by the Director of Navy Culture facilitators. About twenty-five of us sat in a room that felt heavier than any training venue has a right to feel. Sailors are usually loud before a brief. Side comments. Shit talk. Quiet laughs. That room was different. Almost silent. People spoke low and stayed close to those they trusted.

The facilitators tried to lift the energy. You could see the effort. They could also see the fatigue written across the room. You don't need a survey to know when a unit is cooked. You only need to stand in front of them for a couple of minutes.

After a few warm-up exercises, the facilitators moved into the substance of the day. Before the drawings, there were group discussions, the standard culture-workshop questions. What good leadership looks like. What ineffective leadership does to a team. What psychological safety feels like when it exists, and how obvious it is when it doesn't.

People spoke carefully at first, but the answers came quickly. Too quickly. Trust. Consistency. Accountability. Approachability. Listening before deciding. Then the other side of it. Micromanagement. Fear. Unpredictability. Leaders who don't reflect. Leaders who punish honesty. Nothing controversial. Nothing extreme. Just the basics of not fucking people over.

The OIC was right there with us. Contributing. Nodding. Saying the correct things at the correct moments. On paper, she sounded like someone who understood leadership. Someone who could describe exactly what made a team healthy, and exactly what broke one.

That was the unsettling part. She understood leadership in theory. She knew the language, the frameworks, the failure modes. She could describe healthy leadership clearly. She just couldn't see herself anywhere in the picture.

After that, the facilitators moved us into the next exercise. It was meant to shift things out of language and into something more instinctive.

"Draw a ship that represents FLSE-D and its culture," they said. "Use whatever imagery or symbolism feels accurate."

It sounded harmless. A creative exercise. Low risk. Something you could finish without thinking too hard about it.

It wasn't.

Once you stop talking and start drawing, there's nowhere to hide. You can't dress it up or steer it. You end up putting down what the place actually feels like, whether you intend to or not.

She went first.

She stood up with a confidence that felt completely disconnected from the room and held her drawing up like she'd just solved the problem for everyone. There was pride in it. No hesitation. No sense of the weight sitting in front of her.

Then she turned it around.

It looked like it belonged in a children's colouring book.

A neat little boat floating on calm blue water. Sunshine overhead. Soft clouds drifting past. A rainbow arcing cleanly across the page.

And then, I swear to fucking God, there were unicorns trotting across the deck like FLSE-D was some magical floating wonderland.

Unicorns. Rainbows. A fantasy that didn't exist anywhere the rest of us were working.

She described FLSE-D as a positive, thriving unit. She spoke about engaged people, strong performance, and mutual support. A place where sailors enjoyed turning up to work. She didn't hesitate or qualify any of it. She spoke as if it were simply true.

It bore no resemblance to the unit we were standing in. The version she described had energy, confidence, and ease. People weren't bracing themselves before speaking or carrying the quiet strain that had become normal for us.

She smiled the whole time she talked.

The rest of the room dropped their eyes to the desks. Looking at her felt like too much.

The smile never left her face.

I scanned the room. Arms folded across chests. Heads down. Jaws set tight. People focused on their own drawings so they didn't have to meet her gaze. A few sailors traded brief looks that didn't need words. The question was obvious enough without being asked.

The facilitators nodded along politely. They had seen the survey data. All twelve categories sat below effective. They knew exactly what this culture looked like. Facilitation is facilitation. You acknowledge, you thank the speaker, and you keep the session moving.

Even so, their posture shifted.

They had not expected unicorns.

In that moment, it was clear she had no idea what was happening inside her own command. There was no calculation behind it. No attempt to spin the room. She wasn't performing or protecting herself. She genuinely believed she was leading a unit that was healthy, engaged, and functioning well.

That belief was the problem.

She sat back down, still smiling, unaware that the room was about to show her a version of FLSE-D she did not recognise.

One by one, the rest of the unit stood and held up their drawings.

One sailor showed a ship listing hard to port, taking on water faster than the crew could bail. Buckets moved, arms worked, but it was never going to be enough.

Another held up a hull riddled with holes, new ones appearing as quickly as the crew tried to seal the old. Every fix was temporary. Nothing held.

Another drew a deck where the crew were scattered and isolated, heads down, not speaking to one another. The ship still floated, but only just. Survival had replaced coordination.

Someone else showed a vessel under a sky swallowed by storm clouds, lightning tearing through the dark while the ship was thrown around without control, pushed wherever the weather decided.

Then someone held up a ship engulfed in flames. Fire tore through the decks while the crew ran in every direction, trying to contain a blaze that refused to stay contained. Every effort slowed it. Nothing stopped it.

The drawings kept coming like that. Heavy. Unfiltered. Honest. What we had been living showed up on paper without explanation or apology.

There were twenty-five sailors in that room. Twenty-five drawings.

Not one of them had a unicorn anywhere on it.

The contrast was unavoidable. Her drawing sat on the table in front of her, bright and untouched, like it belonged somewhere else entirely. Around it were ships taking on water, crews drowning, leaders absent, and sailors grabbing whatever they could just to stay afloat.

By the time the last drawing went up, the air in the room felt thick. No one shifted in their seat. No one spoke.

Then Leading Seaman Howard Dunn stood.

Howard has never softened an edge in his life. He held his drawing up, and the room went completely still.

The ship was split clean down the middle. Bow and stern drifting apart, both already sinking. In the gap between them were people in the water, reaching up, calling out, disappearing beneath the surface. Sharks circled below, waiting.

Above it all, perched on top of an Eiffel Tower of all things, was her.

Smiling.

Waving.

Completely removed from what was happening underneath.

She shot to her feet so fast her chair scraped across the floor. Her face flushed. Her hands shook. Without a word, she walked out of the room.

Crying.

The facilitators froze.

Twenty-four of us stayed where we were. No one shifted. No one spoke. The room settled into a silence that felt deliberate rather than shocked.

Nothing needed smoothing over. No one reached for humour or deflection. The drawing had already said what couldn't be walked back.

It showed a leader detached from the people beneath her, elevated above the damage while others carried the consequences. No briefing pack or survey result had ever made that clearer.

We knew something else as well. She wasn't crying because she understood. She was crying because the version of herself she believed in had just been exposed in front of a room full of witnesses.

For a brief moment, I let myself hope it might change something. That seeing ships on fire, crews drowning, and people going under would force a reckoning. That the tears meant it had finally landed. That I wouldn't have to keep carrying this alone.

It didn't.

What followed wasn't repair or reflection. It was a fucking escalation.

She doubled down.

She returned to the unit with the same authority she'd always held. Nothing about her position changed. No boundary was placed around her. No protection was put in place for the people underneath her. Whatever recommendations existed stayed abstract and toothless. Control remained exactly where it had been.

The behaviour intensified.

Decisions tightened around her. Direction became harder to pin down. Small issues drew disproportionate attention while real problems drifted unresolved. People stopped raising concerns altogether, not because they didn't exist, but because experience had taught them what happened when they did.

The facilitators didn't come back. No follow-up sessions were scheduled. No one checked whether anything had shifted. The day that was meant to surface the truth closed out as an event, not an intervention.

Inside the unit, the adjustment was quiet and immediate. Sailors narrowed their engagement. Conversations shortened. Risk went underground. People worked around her rather than with her, absorbing damage where they could and passing the rest downward.

The culture didn't stabilise. It thinned out. The work still got done, but nothing settled. Every decision carried an edge. Every interaction took more effort than it should have.

I kept turning up and doing the job, because that's what I knew how to do. The unit needed it. The people needed it. But something in me was starting to give way. Not all at once. Not dramatically. Just enough that I noticed it taking longer to switch myself on in the mornings.

The system had seen the problem clearly enough to measure it. It had watched it play out in a room full of witnesses. Then it stepped back and returned control to the same place it had always been.

That was when the damage stopped being contained to the unit.

That was when it started landing in me.

Six

The Exit Plan

Sometimes even to live is an act of courage.   — Seneca

The weeks after the culture session blurred together. Not dramatically, and not in a way that announced itself. It was the quiet tightening you feel long before anything actually fractures. Something inside me was giving way under a sustained load, the kind of strain metal carries before it finally snaps.

That was when the vomiting became routine.

Every morning at 0500 I woke up, looked at my MMPUs (uniforms) folded neatly over the chair, and felt my stomach knot before I even moved. I walked to the bathroom, gripped the sink, and leaned over the toilet while my body emptied itself of whatever it could find. Some mornings it was bile. Other mornings it was dry heaving that left my throat raw and burning. When it stopped, I rinsed my mouth, straightened up, pulled on the uniform, and left the house as if nothing out of the ordinary had happened.

My body had started telling the truth long before I was willing to listen.

The physical slide came quietly. I ate under stress. Sleep fractured. Training dropped away. Looking after myself stopped registering as something worth the effort. I caught my reflection and barely recognised the bloke staring back. The face looked swollen. The jawline had softened. The eyes that used to light up when I smiled sat dull and flat.

Rhiannon saw the change without needing to name it. She didn't say anything out loud, but it was there in the way she looked at me, the pauses that lingered a fraction longer than they used to. I could see the concern even when neither of us spoke about it.

She wasn't wrong. Something in me had gone quiet.

People at work noticed as well.

"Hey mate, you okay?" "You look tired." "You right, Jacquesy?"

Every time, the answers came out the same way. Practised. Automatic.

"Just tired." "Long week." "All good."

None of it was true. Lying took less effort than trying to explain an exhaustion that didn't yet have language.

I'd been in uniform for nineteen years. One more would make it twenty, the number that mattered. Twenty was the marker. The clean ending. Proof that the grind had counted for something. Nineteen felt unfinished, like stepping off the track just short of the line. I couldn't stomach the idea of walking away without hitting that special number.

I clung to the timeline because it gave me something concrete to hold.

On paper, it looked survivable. She was posting out in January 2025. I planned to discharge in August 2025. Six months of overlap. Six months, I told myself I could endure. I repeated it every morning while I buttoned my shirt, tied my boots, and sat in the car trying to slow my breathing before walking into the building.

But I wasn't surviving. I was drowning in plain sight. By early July, the shift was impossible to ignore. Leaving the Navy stopped being the thought. Leaving everything replaced it. I didn't drift into suicidal ideation. I constructed it deliberately, piece by piece. A plan I rehearsed so often it settled into me with a familiarity that felt disturbingly normal.

My orange 2021 GT 5.0 Mustang became the centre of that plan. Neptune plates. V8. The pride of my garage. The symbol of everything I had built outside the Navy. Four hundred and sixty horsepower. Zero to one hundred in under five seconds. A machine built for speed, control, and presence.

I decided I was going to use it to kill myself.

I knew exactly where. Berrimah, just down from Hidden Valley Raceway. A sweeping right hander I'd driven a hundred times. It tightens without warning, with a concrete barrier waiting for anyone who misjudges it. I used to love that corner, the way the Mustang settled and surged as it came alive underneath me.

In those weeks, that corner stopped being a road. It became an exit. Clean. Definitive.

I thought about it constantly. It followed me into the shower, onto the drive to work, and into the early hours of the morning when sleep wouldn't come. The plan never changed. Take the Mustang out alone. Reach that corner. When the moment came, don't correct. Let physics finish what I no longer felt capable of enduring.

It came down to speed, momentum, and impact. A single decision that ended everything else.

The worst part was that it made sense. Mustangs crash all the time. People joke about them not turning for a reason. It would look like an accident, a misjudgement that went wrong. No one would question it. My family wouldn't have to carry the weight of suicide. My record wouldn't end with that word attached to it. It would just be another bloke who pushed his Mustang too hard on the wrong night.

In my head, it felt clean and contained. A quick end that left no questions behind.

I even timed it out. Not a workday. Not after a confrontation with her. Nothing that tied the cause too clearly to what was happening at work. Just a spontaneous drive that went wrong.

That's how methodical it became. Suicidal ideation turned into problem solving, calm and structured, the same way I had been trained to think.

Looking back, that's the part that unsettles me most. I wasn't distressed or emotional. I felt calm, as if I'd finally found a solution to a problem that had been consuming me.

The vomiting eased. The anxiety on the drive to work disappeared. Even the sight of her white Ford Ranger stopped triggering the usual adrenaline spike. All of it faded in the shadow of that imagined crash.

I drove past the corner dozens of times. I slowed down, looked at the barrier, and walked through the line and the angle in my head, testing myself.

Could I do it?

Would I do it?

Some days, the answer was yes.

That's how far I'd been pushed. To the point where destroying the car I loved, the life I'd built, and the family waiting at home didn't feel dramatic or impulsive. It felt logical.

But I didn't do it.

And there are three people I will be grateful to for the rest of my life. Ozzie. Squizzy. Kiarra. Three Navy Chef mates.

I don't think they knew how close I really was. Maybe they saw something in my face. Maybe they felt a shift in my energy they couldn't quite place. They never confronted me. They never staged an intervention. They never asked the question I wouldn't have been able to answer.

They just showed up.

"Dinner tonight?" "Drinks after work?" "Want to catch up?"

Nothing loaded. No therapy. No digging. Just three mates making sure I wasn't alone, even if they had no idea why their presence mattered as much as it did.

Around the same time, two other things gave me somewhere to step outside my own head. Navy football and the Entourage community. Training sessions, games, conversations that weren't about work or survival. Hours where I could disappear into movement, into learning, into being around people who expected effort but not explanations. They didn't fix anything, but they gave me relief. Space to breathe. A reminder that the world was still bigger than the corner I kept circling.

And then there were the FaceTime calls home. Seeing the kids' faces. Hearing their voices. Watching them carry on with their lives while I sat thousands of kilometres away trying not to unravel. Those calls anchored me in ways nothing else could. They reminded me what was real, what still mattered, and what would be left behind if I let go.

But it was the three mates who kept me going day to day. Sitting across from them, laughing at something stupid, talking rubbish, feeling like myself for the first time in days, I found just enough strength to get through the next few. Enough light to push back what had taken over my mornings, my thoughts, and the shape of my world.

They didn't save me with heroics. They kept turning up.

Their consistency mattered more than anything else. A text. A drink. A seat at a table. It reminded me there was a life outside the uniform, outside the toxicity, and outside that corner at Berrimah where my thinking kept circling.

Those small invitations became the rope I held onto. Not because they were dramatic, but because they were reliable. Proof that I still belonged somewhere that wasn't broken.

Even with that support, I was hanging on by threads.

Things worsened after the culture survey in June. When she stormed out of the room in tears, confronted by twenty-five drawings of sinking ships, she didn't reflect or pause. She latched onto a single recommendation in the report, group activities to "improve culture", and twisted it into something no one had asked for.

Instead of speaking to the team or asking what we actually needed, she issued a decree. Mandatory PT. Early mornings. No discussion. No consent. Just control.

She'd pushed PT before, but she never turned up herself, so it never held. This time was different. This time it was about reasserting authority after being publicly confronted with the truth of her leadership.

The team hated it. We were already stretched trying to support the Patrol Boat fleet. Everyone was exhausted. Forcing people into a mandatory activity they didn't want wasn't improving anything. It made the environment tighter and more resentful.

I was in no condition for it. I was vomiting most mornings. I'd just had my gallbladder removed. My body was already failing. The idea of dragging myself onto base at 0730 for a performance of compliance made my skin crawl.

None of that mattered. She didn't adjust, and the divide widened. The toxicity didn't ease. It settled deeper into the unit, into the days, and into me.

Every morning, before I opened the door to that building, I told myself the same lie I'd been clinging to for months. Just six more months. That I could survive anything for six more months.

I had wrapped my entire identity around that belief, even as my body rejected it in ways I could no longer ignore. The truth was already clear, no matter how hard I tried to outrun it.

I wasn't surviving, and I wasn't pushing through. I wasn't holding the line. I was running out of time, and my body had already started the countdown.

Seven

The Breaking Point

Unlimited power is apt to corrupt the minds of those who possess it. —
William Pitt the Elder

On the morning of 8 July 2024, I woke at 0530 to a notification.
The freezer alarm had gone off. We had lost about seven hundred
dollars' worth of food stock overnight. Not a great start to the day, but
nothing unusual for FLSE-D. We were always one equipment failure
away from chaos. I dealt with it, then got ready for a scheduled medical
appointment, a follow up after losing my gallbladder to pancreatitis back
in March.

The timing mattered. I had just returned from nine days away with
Navy Football Federation Australia. PO Gunn and I had both been on
approved absence during that period. I ran Forces Football in Darwin,
an event I had crafted for Navy FFA where Navy took on the United
States Marines on the fourth of July. It was a rare thing. International
engagement done properly, using sport as a bridge rather than a talking
point.

The event landed exactly as intended. Senior officers from both sides
were present. The standard was high and the feedback reflected it. We
took an already solid relationship and lifted it to another level. It was
work I was proud of, delivered while I was slowly coming apart, and it
mattered well beyond the footprint of our small unit.

None of that registered when I got back.

While we were away, the work followed us. Messages from Lieutenant Commander Carrigan kept coming through. Nothing urgent. Nothing that couldn't wait. Routine matters sent simply because she wanted them actioned immediately, regardless of where we were or what we were doing.

Approved absence didn't buy space. It just meant the work arrived through a different channel.

At the same time, we started hearing what was being said back in the workplace. Frustration about us being away. Approved absence framed as inconvenience. Context didn't feature. In her version, not being physically present made you a problem.

I walked into that morning already carrying the cost of trying to operate at two levels at once. Delivering meaningful outcomes outside the unit while absorbing dysfunction inside it. The freezer alarm was just noise. The appointment was the real marker. My body had already paid for what the system refused to see.

It had been warning me for months. Vomiting. Anxiety. Insomnia. I ignored all of it. I kept turning up, enduring, complying. Eventually something inside me quit. My gallbladder failed. Pancreatitis followed. Emergency surgery. Two weeks of recovery leave.

My body tried to protect me. I still ignored it because that's what the Navy trains you to do. Push through. Ignore the damage. Keep moving, even when your body is waving a red flag the size of a fucking helicopter blade.

When I came back, nothing had changed. Same leader. Same toxicity. Same weight pressing down.

There wasn't time to reflect on any of it. There was a freezer failure to deal with. A medical appointment to attend. A business to run on the side. Navy by day, Neptune by night. Two lives stacked on top of each other, both demanding more than I had left. Over the top of that sat a workplace that expected compliance regardless of workload.

At 0630, a message landed in the FLSE-D group chat. PT was now mandatory for everyone. The class was scheduled for an hour later. She had all fucking weekend to send that message and chose to wait until 0630 on the morning itself.

At 0735, before most people had even arrived at work, she followed it with another message in the FLSE-D Signal chat. A demand for written explanations from anyone who didn't attend her mandatory PT session.

I had already told my Divisional Officer, LEUT Glassrock, about the freezer failure and my medical appointment. It didn't matter. At 0735, she demanded a written explanation for my absence from PT. At 0818, another message.

"I have just been informed of the freezer. I want a DIR. The justification can come later."

A Defence Incident Report for the freezer. There was no interest in what had actually happened. No acknowledgement of the workload already on my plate. No recognition that I was on my way to a medical appointment. The priority was paperwork and compliance. Nothing else.

One minute later, at 0819, another message landed.

"Also, CCIR. DMLO SPT (Deputy Maritime Logistics Officer Support) has the lead in progressing this."

A Command Critical Information Report. She was already escalating it up the chain while I was walking toward the health centre, reading the messages as the anxiety tightened in my throat.

The message was clear enough. Expectations first. Pressure always. Explanations later.

By the time I left the office and started heading toward the health centre, the tension had settled into my chest like an anchor. My phone was still buzzing with the aftershock of those messages when I saw her riding toward me on a bike.

In the entire time I'd known her, I'd never seen her on a bike. As far as I can remember, this was the only day it ever happened.

She came straight at me, awkward and unsteady, riding like someone who was not used to riding a bike. She fumbled for the brakes and pulled up short in front of me.

She fixed me with that familiar condescending stare, the one that made you feel like your presence was an inconvenience she was forced to tolerate. Smaller than your own shadow.

She didn't ask a question. She didn't acknowledge the freezer breakdown. She didn't acknowledge that I was on my way to a medical appointment.

She tilted her chin, her eyes flat and cold, and said it. Not as a conversation. Not as guidance. As a certainty.

"I want you in my office when you return."

There was no context and no curiosity. It landed as instruction, not conversation.

The timing didn't matter. Neither did the circumstances.

The expectation was simple. When she said it, I would comply.

My body reacted before I had time to think. Hands shaking. Heart thudding hard against my ribs. A surge of heat and pressure moving through my chest that had nothing to do with anger. It felt closer to humiliation, sharp and immediate, the kind that locks you in place.

She didn't speak to me like a sailor with nineteen years on the job. She spoke as if I were an inconvenience that needed managing.

I stood there and answered, "Yes, Ma'am".

Quiet. Automatic.

The words erupted without thought. Not agreement. Not compliance. Just the response that kept things from getting worse.

I made it to the health centre, but my head wasn't there. I couldn't focus. My breathing stayed shallow and uneven. Her tone kept replaying. The messages. The encounter on the bike. It all looped while something inside me shifted, like a warning light that had been flashing for too long and was about to fail.

I knew I wasn't in a good place. I knew walking into the weekly management meeting at 0930 was a bad idea. Not inconvenient. Not uncomfortable. Dangerous. Something instinctive was telling me I was right on the edge, that one more hit would push me past it.

What still turns my stomach is this. I was already at the health centre. I was standing in the very fucking building where you're meant to admit you're not okay. Any sensible person would have stopped there. Said something. Asked for help.

Not me. By then, I didn't trust myself to stop.

So I did what the Navy had drilled into me over nineteen years. I swallowed it and kept moving, even though a part of me understood with complete clarity that walking into that meeting was a mistake I wouldn't be able to undo.

By the time I stepped into the room at 0944, I was late and spent. The residue of the encounter outside clung to me, heat sitting high in my chest and neck. I knew my face gave me away. Hollowed out. Worn thin. Held together by habit rather than strength.

I took my usual seat, the same chair I'd parked my arse in for two and a half years, directly opposite her at the far end of the table. I slumped into it, shoulders heavy, hands trembling so I tucked them under the table where no one could see.

I knew I looked like a bag of shit. I felt like a bag of shit.

The room had that sterile Navy feel. Cold air conditioning humming. Fluorescent lights buzzing overhead. No windows. A long conference table with one door in and out. Papers spread everywhere. Pens lined up. Half-drunk coffee cups left behind. A typical Monday mess.

She was talking, wrapping up her expectations for the week. It was her usual routine. She went first, laid out her plan as if it were fixed, and only after that, if there was time, she opened the floor to the team.

By then, the shape of the week was set.

Finally, she asked, "What does everyone have on this week?"

We all knew the script.

"Business as usual, Ma'am."

Not because it was true. Because saying anything else never changed the outcome. By the time the question was asked, the decisions were already made.

Around the table sat the deputy logistics officers, Finance, the Chief Reg, a couple of others. No one met her eyes. A pen rolled back and forth between someone's fingers. Jacinta shifted in her seat, shoulders tightening like she was bracing for impact.

Everyone felt it.

Something was coming.

We just didn't know what form it would take.

The discussion moved to mandatory PT. She said it was necessary for team connection.

Jacinta spoke up. Her voice stayed steady, but the frustration sat just under the surface.

"Ma'am, we already do a lot of welfare activity every week. Is the priority welfare, or supporting the Patrol Boat fleet?"

The reply came straight back.

"That's fine, Jacinta. I'll just cancel all other activities."

Jacinta didn't take the bait. "Ma'am, that's not what I'm saying. I'm pointing out we're already out of the office a lot, and compulsory PT doesn't work for every depart..."

SLAM

Her book hit the table.

The crack cut through the room. Sharp and sudden. Pens stopped moving. Shoulders tightened. Every head rose at once.

She leaned forward, eyes locked on Jacinta, jaw set hard.

"I don't give a fuck who my decisions upset. This is what I want, and that's final."

The room dropped into silence. Not awkward. Not polite. Heavy enough to feel.

For a moment, everything around me dulled like a detached float that hits just before your nervous system starts pulling power. The edges softened. Sound thinned out. My vision narrowed until she was the only thing left in focus.

She sat back in her chair, shoulders squared, chin lifted. There was a small smile at the corner of her mouth. Not accidental. Not nervous.

It was the look of someone satisfied with what they had just thrown into the room.

The sentence looped in my head: *"I don't give a fuck who my decisions upset."*

She hadn't softened it. She hadn't framed it. She hadn't bothered to dress it up as necessity or trade-off. She said it plainly, in front of her management team and her senior sailors, the people keeping FLSE-D functioning day to day.

The message didn't need explanation. What mattered was settled. What we carried didn't register. What counted was what she wanted.

Everything I believed about leadership collided with what was sitting across the table.

Something inside me stopped negotiating.

My heart wasn't racing. It was hammering, heavy and relentless, like a fist driving into my chest. My hands shook under the table. Heat crept up my neck as my breathing tightened. Fourteen months of pressure compressed into a single moment. Weeks of vomiting every morning. A uniform that felt heavier each day. The lie I'd been telling my family that I was fine, even as I was falling apart.

It all compressed into that moment. The pressure reached a point it could no longer hold. Nothing snapped; it gave way.

I felt the shift as it happened. The point where your mind stops trying to explain or justify and switches instead to self-preservation. The quiet understanding that you've crossed a line of no return.

I knew what I was about to do. I understood the cost and the consequences. In the military, you don't challenge a superior officer. Not publicly. Not in front of witnesses. Not like this.

My body moved before I had time to talk myself out of it. I pushed my chair back and stood. The legs scraped across the floor, a long, sharp sound that cut through the room and pulled every head toward me.

I met her eyes and didn't look away.

My voice wasn't loud. It wasn't steady either. It carried the strain of months of holding my tongue and swallowing my self-worth. When I spoke, there was no room left to soften it.

"With all due respect, Ma'am," I said, "you'll need to find another PO Chef, because I can no longer work for a dictator."

No one spoke. No one moved. It felt like everyone had forgotten how to breathe.

Let me be clear about what I had just done.

In the Navy, this isn't frustration. It isn't a disagreement or someone having a bad day. It's career suicide. Public. Deliberate. Irreversible. Nineteen years in uniform. Months from twenty. A family at home. A reputation built slowly, the hard way. And I'd just put all of it on the line in one sentence.

I couldn't do it anymore.

I couldn't sit through another meeting where my voice carried no weight. I couldn't keep absorbing days where my people didn't matter, run by someone who treated the unit like equipment. Used until it failed, then blamed for failing.

So I said it. Out loud. In a room full of officers and senior sailors. No whispering. No backtracking. No attempt to soften the edge.

"I can no longer work for a dictator."

Everyone in that room understood exactly what had happened. I'd ended my career.

I turned and walked toward the door, knowing I could have stopped there. Even that alone was enough to bury me under the Defence Force Discipline Act.

But something inside me, the last part that hadn't been smothered, made me stop.

I turned back.

She sat there motionless, smaller than I'd ever seen her, eyes fixed on the table like she might disappear if she didn't look up. No one moved. No one interrupted. Something in me refused to let it end with silence.

I stepped forward just far enough that she had no choice but to lift her eyes.

When she did, I gave her the truth she'd spent fourteen months avoiding.

"I thought we were supposed to be a team, not a dictatorship."

I said it twice.

The first time came out as instinct. The truth breaking free after months of being held down just to survive. The second time was deliberate. I wanted her to hear it again. I wanted everyone in that room to hear it, with no place for reinterpretation or retreat.

Then I turned and walked toward the door.

I don't remember exactly what happened next. Not the movement itself. Not the sound. I know it happened because enough people told me afterward, and because the impact travelled farther than I did.

The door slammed hard enough to be heard throughout the building. The frame shook. It bounced back open. People in the corridor stopped mid-step.

In nineteen years in the Navy, I had never slammed a door. You don't do that. You don't storm out of meetings. You don't make noise on the way out, especially not after calling your OIC a dictator.

But that door echoed.

Not as a performance and not as a threat. It happened at the point where restraint finally ran out. Where control, optics, and silence stopped mattering.

I was done.

And I didn't look back.

Eight

The Aftermath

Trauma is not what happens to you. It is what happens inside you as a result of what happened to you.   — Gabor Maté

After I left that room, I didn't wander.

I walked straight to my desk. Fifteen metres from the door. Close enough that the sound of it still sat in my ears. Close enough that nothing had time to cool or settle.

I grabbed my bag without thinking. No sorting. No checking what was inside. Just muscle memory doing what it needed to do to get me out.

LSML-C Bendetto looked up as I moved. He didn't ask what happened. He didn't need to. Whatever was on my face told him enough.

"You're in charge," I said. "I'm fucking out."

That was it. There was no pause to brief, no space to explain, no attempt to smooth anything over. The handover happened in a single sentence because it had to. The work couldn't be left unattended, and I wasn't staying. Someone needed to be holding the line the moment I stepped away, and Bendetto was standing there when the decision landed.

I didn't slow down or look back. There was nothing to reconsider and no cleaner version waiting to be found. What was said had landed, and the moment was already closed. I turned and headed straight for the office of the Executive Officer.

My body knew where it was going before my head caught up.

Out of the FLSE-D shed, past the freezers the team were still working on, up the staircase and across to the new Command Building. Through the doors, up the stairs, straight into the XO's office. I didn't hesitate at any point. Muscle memory took over and carried me the whole way.

Commander Clair langham was there. Lieutenant Commander Bunton was with her. Both of them looked up the moment I came through the door.

I don't remember what my face looked like. I remember the instant they saw it. Whatever was sitting on me arrived before any words could. I didn't offer a greeting or rank. I didn't lead with anything at all.

That was the moment it gave way.

I broke.

Not the tidy version people imagine when they talk about breaking. Not something you hide in a bathroom or swallow until you're alone. This was a full systems failure. My chest seized, my throat closed, and the sound came out of me before I could stop it. Deep, violent sobs that folded me forward and stole my breath. The kind that don't ask permission and don't care who's watching.

There was no dignity left to protect. No professional mask to hold in place. I wasn't managing anything anymore.

Weeks of swallowing it. Months of pretending everything was fine. Vomiting every morning before work and telling myself it was just stress. Losing my gallbladder because my body finally had enough. Being told my explanations didn't matter. Being told to comply, endure, and keep producing regardless of the cost.

Whatever reserve I'd been running on finally emptied.

"Sit down," Commander Langham said.

It wasn't sharp and it wasn't soft. It was command voice used properly. Clear. Anchoring. Non-negotiable.

I don't remember walking to the chair. I remember my hands shaking so badly I couldn't steady them. My breathing stayed high and shallow, like my body had forgotten how to come back down. There was no attempt to regain control because there was nothing left to control.

I cried like someone whose nervous system had finally hit the wall. Loud. Ugly. Uncontained.

A few minutes later, PO Jacinta Gunn arrived.

She took one look at me and crossed the room without hesitation. She wrapped her arms around me and held me there. Solid. Grounded. Human.

That contact cut through the spiral just enough to stop me coming apart completely. It didn't fix anything. It just slowed the fall.

"Are you okay?" she asked quietly.

I wasn't. But for the first time in fourteen months, I didn't have to lie about it.

We stayed in that office for the next ninety minutes. Me. Jacinta. The XO. Lieutenant Commander Bunton.

There was no ranting and no theatrics. No emotional dumping for effect. Once the worst of the physical collapse eased, what followed was controlled and deliberate. Almost clinical.

We went through it from the beginning.

The culture survey and what it actually showed once the numbers were stripped of spin. Trust eroded. Psychological safety absent. People operating in self-protection mode because it was the only way to get through the day without becoming the next target.

The Vision, Mission, and Values work. How it started as something built by the unit and was slowly stripped of the people who created it. Ownership removed decision by decision until nothing of the original intent remained.

Mandatory PT imposed as control rather than care. Welfare reframed as compliance. The underlying message that bodies were there to be directed, not listened to.

The freezer incident. The timing. The messages that morning. The expectation that everything be actioned immediately regardless of context. Emails sent without explanation. Decisions made without understanding the work underneath them.

Pressure applied without limit.

Then the physical cost. The anxiety. The constant sense of threat walking into the building. The gallbladder. The recovery that didn't actually allow recovery because nothing changed when I returned.

All of it. Laid out cleanly. In order. No exaggeration required.

Commander Langham listened.

She didn't interrupt to reframe it. She didn't minimise it. She didn't defend the OIC. She didn't reach for easy explanations like miscommunication or personality conflict. She took notes. She asked questions when something needed clarity. When it didn't, she stayed quiet.

That mattered more than reassurance ever could.

When we finished, the room went still. Not awkward. Settled. Like something heavy had finally been put down.

She looked at me and said, "You need to take leave. You need to step away from this and recover."

It wasn't advice, and it wasn't framed as a suggestion. It was direction.

For the first time in fourteen months, someone in leadership acknowledged the cost. Not the optics. Not the inconvenience. Someone recognised this wasn't a bad day or a difficult interaction. It was accumulated damage. Someone told me it was acceptable to stop.

For a brief moment, I believed that was the line. That something had finally shifted. That the system had recognised what it had allowed to happen.

I had just broken down in front of the Executive Officer. Everything was on the table. I'd been directed to take leave and recover.

It didn't even last three hours.

At 1215 that same day, my phone buzzed.

A text message.

From her.

"I have just been notified of your intentions to take leave. This is not approved noting the issues currently being encountered by your team at this present time. You need to step up and take charge.

Further, you are to report to me tomorrow first thing so we can discuss your outburst and a way forward.

Your behaviour this morning was unacceptable for a Petty Officer.

I expect to see you in my office at 0830."

I read it once. Then again. Then I just stared at the screen.

The direction from the Executive Officer had been ignored. My breakdown had been reframed as an outburst. The history, the surgery, the context, the timing all disappeared.

What remained was expectation.

The instruction was simple. Report. Explain yourself. Resume control.

There was no question about my wellbeing. No acknowledgement of the conversation that had just taken place. No pause to consider what pushing again might do. The expectation was compliance.

I didn't reply.

Silence was the only thing I had left that hadn't already been taken.

When I didn't respond, she escalated.

The Deputy Maritime Logistics Officer, SBLT Ruben, called me.

"If you don't show up tomorrow, you'll be listed as AWOL."

Absent Without Leave. A chargeable offence. Delivered without emotion, like it was just another administrative step.

The message was unmistakable. Ignore the XO. Ignore medical reality. Ignore what had just happened. Report anyway.

I didn't.

The next morning, I reported to the health centre instead.

In nineteen years of service, I had never done that. I had never walked into the health centre and asked for mental health help. I'd always managed it myself. Absorbed it. Told myself it would settle if I just held on a bit longer.

This time, I didn't try to manage it.

I told the doctor everything. From the start. No filtering and no minimising.

The diagnosis came quickly. Acute workplace stress.

Seven days of mental health leave was issued immediately.

Two weeks later, after further assessment, the formal diagnosis followed. Acute Mental Health Disorder.

I spent the next three months in treatment.

For the first three weeks, I cried constantly. Every time I spoke about what had happened, the same physical response took over. My chest tightened. My hands shook. My breath locked high in my throat. The words wouldn't come out without everything else spilling with them.

Doctors. Psychologists. Friends. Each retelling reopened it. My body reacted before my mind could organise the story.

That's when I understood how deep it had gone.

This wasn't frustration or anger. It wasn't exhaustion or burnout.

This was injury.

The triggers were everywhere. Every time I saw a white Ford Ranger, my body reacted before I had time to think. Palms sweating. Muscles locking. Breathing cutting out mid-cycle. I'd have to pull over, sit there, and wait for my nervous system to stand down. It got so bad I banned white Ford Rangers at Neptune.

That reaction didn't care that I was safe. It didn't care that she wasn't there. My body had learned something, and it no longer listened to logic.

And I wasn't the only one.

Three other people from FLSE-D were placed on mental health leave within the same twelve-month period. I won't name them. Their experiences aren't mine to tell. But the pattern matters.

Four people. One unit. One leader.

That isn't bad luck. It isn't coincidence.

It's impact.

What followed was procedural. Slow. Impersonal. Predictable.

I was sent to an Individual Welfare Board. An IWB. Navy's mechanism for dealing with situations like this. The right people in the room. Medical. Chain of command. Support staff. All of it structured, formal, and recorded.

I told the story again.

Partway through, the doctor stopped me. Not for clarification. To point something out. When the conversation returned to the OIC, my body reacted immediately. Breathing tightened. Hands shook. Posture changed. The response was visible before I said a word.

The board noted it.

The outcome was temporary containment, not resolution. I was posted to Hospitality and Catering (H&C) for four weeks, removed from the environment that had done the damage. After that, I proceeded on Long Service Leave for most of the remainder of 2024.

And fuck me, I needed it.

I stayed away. I recovered enough to function. Not enough to forget.

What shifted most wasn't my view of her. That had already settled. What changed was my understanding of the system around her.

The system had the data. It had the survey. It had the stories. It had watched people break. And when the moment came to act, it defaulted exactly as it was built to.

Process over protection.

This isn't a chapter about rage or revenge. It's about aftermath. About what happens once the doors close, the voices quieten, and the paperwork takes over.

I didn't walk away from that experience stronger. I walked away injured. Altered. Carrying something that hadn't existed before.

Before any lessons could be extracted, before anything useful could be built from it, there was one more thing to face.

Not a person.

The system itself.

Nine

When the System Responds

The purpose of a system is what it does.   — Stafford Beer

On 10 July 2024, one day after being diagnosed with workplace stress, I did something I never thought I would do in uniform.

I filed an official complaint against my Officer-in-Charge.

It wasn't dramatic. There was no surge of adrenaline or sense of defiance. It felt procedural. Necessary. Like finally applying pressure to something that had been leaking for too long.

I sat at my desk and opened a blank document.

The diagnosis hadn't fixed anything, but it had given me language. It gave shape to what had been happening inside my body and head for months. The complaint wasn't written in anger. It was written because silence had stopped being survivable.

I started at the beginning and worked forward.

Every incident. Every decision. Every pattern of behaviour that, on its own, could be explained away, but together formed something coherent and dangerous. I didn't embellish. I didn't speculate. I wrote only what I had seen, experienced, or witnessed directly.

When it was finished, it surprised me how much there was. Not because I was trying to make a point, but because the pattern ran deeper than I realised when it was finally laid out in order.

This is what I submitted. It has been slightly condensed for this book, but the substance is exact.

Subject: *Official Complaint Against OIC of Fleet Logistic Support Element-Darwin (FLSE-D)*

Dear Sir,

I am writing to formally raise a complaint against Lieutenant Commander, the Officer-in-Charge (OIC) of Fleet Logistic Support Element-Darwin (FLSE-D). My concerns pertain to the workplace culture at FLSE-D and specific incidents that have affected my trust and ability to function effectively in my role.

Workplace Culture at FLSE-D *The workplace culture at FLSE-D is notably tense, particularly when the OIC is present. Personnel feel they are walking on eggshells. In her absence, the atmosphere noticeably shifts to one of calmness and ease.*

Handling of Petty Officer Joel Howard' Situation *A high-performing senior sailor, PO Howard was removed from the unit shortly after raising complaints about issues at FLSE-D. The OIC contacted previous divisional officers seeking negative information about him, seemingly to further undermine him. PO Howard has raised psychosocial workplace hazards regarding the current culture, which I believe warrants serious consideration.*

Disparate Treatment of Sailors *I have witnessed the OIC's inconsistent handling of workplace issues involving two sailors. One was treated with respect and empathy; the other with anger and disdain. The male sailor, who was the victim, received the harsher treatment.*

Unprofessional Conduct During Meetings *During an N4 (Head of Logistics Division) handover in Darwin, a meeting with Cape Capricorn* *command was undermined when the OIC entered and hijacked the discussion in front of senior officers, damaging established goodwill.*

PT Mandate and Communication Issues *On Monday 8 July at 0735, the OIC demanded written explanations for non-attendance at mandatory PT. I had already informed my Divisional Officer, LEUT Glassrock, of a freezer issue and my medical appointment. During a management meeting, when concerns were raised about enforced PT, the OIC threatened to remove welfare activities and stated she did not care who she upset and that her decisions were final.*

Impact on Mental Health *The cumulative stress from these incidents has severely impacted my mental health. I am anxious about attending work and have struggled to sleep due to workplace stress. In nearly nineteen years of service, I have never previously sought mental health support.*

Refusal to Approve Leave *After discussing my concerns with the XO, I received a message from the OIC stating my leave would not be approved and directing me to report to her office. She instructed SBLT Ruben to inform me that failure to appear would result in being listed as AWOL.*

Conclusion *I am formally requesting an investigation into the OIC's conduct and the overall workplace culture at FLSE-D.*

Sincerely, PO Jacques

I hit send.

Everything now existed on record, not as accusation or emotion, but as a formal account of what had happened and what it had cost.

In nineteen years of service, I had never filed a complaint against a superior officer. I'd absorbed pressure, worked around bad leadership, and told myself that endurance was part of the job. That you keep your head down, protect your people where you can, and don't make things worse by speaking up.

This time, silence felt like participation. Staying quiet meant accepting that this was how people would continue to be treated, and I wasn't prepared to carry that.

So I submitted it.

Not long after, I was interviewed by the Fact Finding Officer. I told the story again, from the beginning, walking through incidents I'd already replayed a hundred times in my own head. It was exhausting. I knew other people were being interviewed as well, but I had no idea what they'd said, how it had landed, or how much of it would survive the translation into official language.

That uncertainty ate at me. The waiting created space for doubt to creep in. Had the last fourteen months actually happened the way I remembered them, or was I losing perspective? Was this a pattern anyone else could see, or was I just the one who finally broke under it?

Once the statement was given, there was nothing left to do.

I waited.

And waiting became its own form of punishment.

Days slid into weeks, then into months. Life didn't pause for investigations. Treatment appointments, paperwork, recovery, and day-to-day admin filled the space while the process ran somewhere out of sight.

Most of the time, it was quiet. Emails came through, but they were usually routine. Broadcasts. Policy updates. Administrative noise that meant nothing until, suddenly, it might mean everything. Every so often I'd open my inbox and feel that brief flicker of expectation. Not panic. Not dread. Just the question that never quite left: is this it, or is it still going?

Most days, it wasn't.

During that period, I was still in treatment. That continued through to November. Appointments. Assessments. Slow work on getting my nervous system back under some kind of control. I was honest about what my body was doing and how the environment had affected me. What I didn't say out loud was how close I'd come to stepping away from everything altogether.

That wasn't denial. It was calculation.

In Defence, you learn quickly what certain words cost you. No matter how often people say the stigma is gone, you see what happens when someone says the wrong thing the wrong way. Doors close quietly. Opportunities disappear without explanation. Your name starts carrying weight you didn't put there. I'd seen it often enough to know exactly where that line sat.

So I spoke carefully. I stayed inside language that would get me help without burning everything else down. Treatment continued. Recovery came in fragments. And behind all of it, the investigation kept moving, silent and distant.

That uncertainty was corrosive.

Some days, I convinced myself the system would work. That the culture survey, the medical diagnoses, the witness statements, and the timelines would add up to something that couldn't be ignored. That enough evidence, laid out cleanly, would force action.

Other days, I remembered every story I'd ever heard about complaints that went nowhere. Officers quietly moved sideways. Sailors told to be more resilient, to harden up, to get on with it. On those days, expectation felt naïve, and hope felt like a mistake.

The investigation stretched on until November 2024.

When the outcome finally arrived, it didn't come with a conversation or a phone call. There was no briefing, no context, no acknowledgement of what had preceded it.

It came as an email.

Just a message sitting in my inbox, easy to miss if you weren't looking for it.

One sentence carried the weight of the whole thing.

"Based on the evidence, it is my decision that the allegations of unacceptable behaviour by the OIC are unsubstantiated."

I read it once. Then again. Then I stopped reading.

Four people had been placed on mental health leave within twelve months. There were formal diagnoses and months of treatment. A culture survey showing a unit in clear decline. Witness statements. Timelines. Medical intervention. Executive Officer involvement. An explicit threat of AWOL.

All of that existed on record.

And the conclusion was unsubstantiated.

It read as if none of it had happened. As if the damage didn't exist because it couldn't be shaped into something the system was prepared to recognise.

I didn't feel angry.

That surprised me.

There was no disbelief. No urge to push back. No impulse to draft a response or reread the sentence looking for an angle I'd missed. I understood the decision before my eyes finished the line.

My body didn't react.

And that was what unsettled me most.

I felt nothing.

Just a hollow stillness, like something essential had been switched off without ceremony. The numbness that arrives after pain has burned through everything it can reach and finds nothing left to touch.

That was the outcome.

Just a return to routine, as if the last year of our lives had been nothing more than an administrative inconvenience.

Business as fucking usual.

Sitting there with the email open, it became clear that nothing had gone wrong. The system hadn't missed anything or failed to act. It had responded exactly as it was designed to respond. Information went in, process ran its course, and hierarchy closed over itself.

What mattered wasn't the people inside it or the damage that had been done. What mattered was containment.

If four people breaking under the same leader, supported by medical diagnosis, formal process, witness statements, and time, wasn't enough to trigger accountability, then there was no threshold that would ever be met. The absence of response wasn't a delay or an oversight. It was the conclusion.

I closed the email and sat back in my chair.

There was nothing left to wait for.

This wasn't closure or justice, and it didn't feel like disappointment. It was confirmation. The system wasn't built to protect people like me, and it never had been.

Once that settles, you stop expecting rescue. You stop waiting for reform. You stop offering yourself up as evidence.

You simply step away.

Ten

Redemption

Circumstance does not make the man; it reveals him to himself. — James Allen

While the institution took months to decide that nothing had gone wrong, my own timeline kept moving. Recovery, fallout, and opportunity ran side by side, overlapping in ways that only make sense in hindsight. There was no clean sequence. No pause while one thing finished before the next began.

Before November arrived and the system shut its door, something else was already in motion.

While I was posted across to H&C at HMAS *Coonawarra*, the days settled into something almost functional. I worked under CPO Matty Cole. The expectations were clear. The environment was steady. I wasn't managed through pressure or performance. I was allowed to do the work without bracing for impact.

It was exactly where I needed to land.

Matty let me do the work without interference. The space allowed thinking again. I wasn't fixed, but I was stabilising. For the first time in months, my body stopped reacting every time I walked through the door. Recovery didn't come from rest alone. It came from being in an environment that didn't require constant self-protection. I'd assumed that was as far as things would go.

My discharge was already in motion. After July, I'd written off anything beyond getting through the remainder of my time intact. Promotion and progression felt finished. Not taken from me. Just no longer relevant.

Then, in August 2024, the Career Manager (Poster) in Canberra called.

"We'd like to offer you a promotion to Chief Petty Officer, effective January 2025."

I didn't answer straight away.

Barely three weeks earlier, I'd come apart; I was still in treatment. Still processing what had happened. Still waiting on the outcome of an investigation. My focus was getting through the three weeks before long service leave.

And yet, there it was. An offer that assumed continuity. Capability. A future I'd already started to step away from.

I didn't hesitate.

"Absolutely."

On the surface, it didn't make sense. What had just happened and what was now being offered existed in the same timeline, without acknowledging each other.

The investigation wasn't about me. It never was. It examined her conduct, based on a formal complaint I'd lodged. While that process ran, nothing else stopped moving.

My promotion paperwork had been submitted months earlier, assessed on work completed well before July, and was already moving through the system. By the time the investigation concluded in November, the decision had already been made. Nothing was paused. Nothing was revisited.

I didn't get promoted because of what happened in July. It happened alongside it.

People talk about functioning alcoholics. I was a functioning but broken man.

The promotion had nothing to do with the breakdown. It came from everything that existed before it. Work done quietly and consistently. Standards held when it would have been easier not to. People looked after. Problems solved without needing credit.

Even at my worst, while everything else was coming apart, that foundation was still there.

The system did not reward the collapse. It recognised the record that already existed.

I needed something to pull me forward. Something to rebuild myself around. Something to prove that what happened to me wasn't the end of my story.

So I took the promotion.

Not because I felt ready. I was still patching myself back together.

I took it because I needed purpose. Something forward-facing. Something that reminded me I wasn't finished yet.

Shortly after accepting the promotion offer, I proceeded on long service leave. I headed back to Cairns to spend time with Rhiannon and the kids. After everything that had happened, that time mattered more than I can properly explain. It wasn't a reset or a fix, but it reminded me of who I was outside the uniform and why I was still here.

On 8 September, I flew to Daydream Island for a week with the Entourage. It was my first in-person experience with the group. Earlier in this book, I said those people saved my life, and I meant it. That week proved it.

Daydream didn't feel like an escape from reality. It felt like being welcomed back into it. The energy, the conversations, the absence of judgement — all of it hit at exactly the right moment. I arrived carrying more than anyone there knew, and I was accepted immediately, without explanation or qualification. That alone was powerful.

There are four people from that week who need to be named. Belle. Matty V. Nick H. Henry. Each of you had an impact on me that went far beyond what you'll ever be aware of. You showed up exactly as you were, and that mattered more than advice ever could.

And then there was Adam. Entrepreneur. Father. Husband. Mate. That first encounter at the bar is burned into my memory for a reason. We'd just been told to dress up. You were standing there in thongs. I walked over, introduced myself, and said, "You look like my type of bloke."

It was one of those moments that reminds you what real connection looks like when there's nothing to gain from it and nothing being performed.

To the Entourage staff; Shane, Stev, Johny, and Jack — thank you. That week stands as one of the best of my life. Not because it fixed me, but because it stabilised me. What you taught, modelled, and reinforced during those days set the conditions for everything that came next.

Before that course, I spent three weeks in October with Navy Football, then transitioned straight into the Chief Petty Officer Leadership and Management Course.

The course covered leadership theory, management principles, strategic planning, decision-making under pressure, team dynamics, conflict resolution. All important. All familiar. Not just from the Navy, but from seventeen years of building and running businesses alongside it.

Much of what was being taught wasn't new to me. It was language for things I'd already been doing.

So instead of sitting back and absorbing content, I found myself contributing. Sharing practical approaches. Breaking concepts down into real-world application. Drawing on scenarios from business and service where decisions had consequences and teams either worked or didn't.

The frameworks I talked through weren't theoretical. They were built from experience. Systems for building teams that trust each other, managing conflict before it turns corrosive, holding people accountable without breaking them, and creating environments where people actually want to turn up and perform.

I didn't recognise it at the time, but I was already teaching the foundations of what would later become The Operator System. The framework didn't exist yet, at least not in name, but the principles were there. They had been shaped over years of leading teams in both military and business environments, refined through success, failure, and pressure.

People noticed. Not because I was trying to stand out, but because I spoke about leadership from experience rather than theory. I wasn't repeating doctrine. I was making it usable.

Every assignment, presentation, and discussion carried that history. I brought what I'd learned from service, from entrepreneurship, and from the last three months in particular. From FLSE-D, I carried a clear understanding of what toxic leadership does to people and what it costs to survive it.

What I shared practical, not abstract. How trust is built and how it erodes. How teams fracture under fear. How leadership decisions keep echoing through people long after the meeting ends. And how, when it's done properly, leadership stabilises, giving people room to perform rather than bracing for impact.

I showed up every single day with purpose. Not chasing recognition. Not trying to control the room. Just doing the work, contributing where I could, and leading when it mattered. What I needed wasn't validation from others. I needed to know, for myself, that I was ready.

Graduation day came and went the way these things do. A room full of uniforms. Family members lining the back wall. Instructors and staff standing off to the side. Names called. Certificates handed over. Polite applause after each one.

I wasn't thinking about awards. I was looking forward to it being done.

Near the end, the Commanding Officer paused and picked up another certificate.

"And finally," he said, "the award for Dux of the Chief Petty Officer Leadership and Management Course."

He read the name.

"Petty Officer Matthew Jacques."

It took a second to register. Then I stood and walked forward. No rush. No hesitation. Just steady.

The CO shook my hand. "Congratulations," he said.

"Thank you, sir," I replied.

I turned to face my classmates and instructors. The room was quiet.

"Thank you to all of my classmates and instructors," I said. "I said on day one that I live by a rule, and that rule is to surround myself with people who are smarter than me. And that's all of you. This award is for all of us. It might have my name on it, but I couldn't achieve it without all of you. So thank you."

The applause happened, but it wasn't the point. What mattered was the shift that followed. A steadiness I hadn't felt in a long time.

I didn't need the award to explain the months that came before it, and I didn't need it to justify anything. It didn't erase what had happened.

It marked something simpler than that.

I was still standing, and I could still lead.

When I returned to HMAS *Coonawarra* in January 2025, I drove through those gates as Chief Petty Officer Matthew Jacques, H&C Regulator.

The same gates I'd driven through six months earlier in tears. The same base where I'd been worn down and finally cracked.

But I wasn't the same man who'd left.

What happened didn't make me better. It made me clearer. Clear on what toxic leadership does to people. Clear on what happens when power goes unchecked. Clear on what leaders owe the people under them.

I came back with a line I will never cross and a standard I will never compromise. No one under my charge will ever be made to feel small, disposable, or alone the way I was.

This wasn't redemption.

It was me taking myself back.

Eleven

The Turning Point

We do not rise to the level of our expectations. We fall to the level of our training. — Archilochus

On 6 December 2024, I was promoted to Chief Petty Officer.

The ceremony was clean, familiar, and brief. The rank slid onto my shoulder the way it always does, practised, procedural, as if nothing significant had changed at all.

Less than two weeks later, I was lying in a hospital bed in Cairns, bleeding internally and far closer to death than I understood at the time.

My return to HMAS *Coonawarra* didn't begin at the base. It didn't start with a briefing, a handover, or a walk through the gates in a new rank. It began there, on that hospital bed.

Before I drove back through those gates in January 2025 as a Chief Petty Officer, my body failed in a way I could no longer ignore. Not because of leadership. Not because of stress. Because of me.

Years of pushing through fatigue. Ignoring warning signs. Delaying recovery. Telling myself I'd deal with it later while my body kept asking for now. In December 2024, it stopped asking.

Before I could lead anyone again, my body forced a reckoning. It was not gradual or subtle, and it left very little untouched. What happened next set the conditions for everything that followed.

I was home with Rhiannon and the kids on Christmas leave. For the first time in months, I felt steady. Capable. Like I might finally be finding my way back. The course was behind me, the future felt possible, and I allowed myself to believe, quietly, that the worst of it was over.

I had no idea what was coming.

It started as stomach pain, then vomiting. Sharp and twisting, bad enough to notice but not yet bad enough to scare me. I told myself it would pass.

It didn't.

By Friday the 20th, I finally gave in and went to the doctor. He reviewed the X-ray, barely paused, and told me I was constipated. He joked that I was more backed up than the Woolworths car park in Darwin, wrote a script for laxatives, and sent me home as if that explained everything. I followed the advice, and on Saturday morning it finally passed.

Rhiannon took the kids up to Atherton for the day so I could rest, and I stayed home, exhausted and weak, assuming my body just needed time to recover. I lay down expecting sleep to do its job. Instead, the pain arrived in full force. It wasn't discomfort or cramps. It was sharp, deep, and overwhelming, the kind of pain that doesn't fade or settle, the kind that tells you something is seriously wrong.

The pain surged hard and fast, sharp enough to cut through thought. It dropped me to the floor before I could react, my body folding in on itself as I struggled to breathe. I lay there curled up, gasping, trying to ride it out, until I managed to grab my phone and message Rhiannon that I might need to go to the hospital.

I hadn't eaten properly since Monday, only small bites when I could tolerate them. By the time she got home, the painkillers I'd taken earlier had dragged me under. I was slumped on the couch, barely responsive, unable to move. My body felt heavy and uncooperative, like systems were shutting down one by one and I was no longer in control of any of them.

I didn't move all Sunday. I stayed where I was, drifting in and out, aware enough to know something was wrong but too drained to do anything about it.

What frightened me most wasn't collapsing earlier, or even the pain that felt like my insides were tearing themselves apart. It came later that afternoon, when I tried to eat something, anything at all, and realised nothing tasted the way it should. It wasn't off. It wasn't spoiled.

Everything tasted like metal.

Like chewing on a fucking spoon.

That was the moment instinct finally cut through stubbornness. This wasn't constipation, and it wasn't something I could grit my teeth through or sleep off. Something inside me was seriously wrong, and for the first time, I stopped trying to minimise it.

On Monday morning, I still had one thing I needed to do. I drove our eldest daughter, Addison, back to her mum's place. I had just enough strength to get through the hour-long drive, focusing on the road and nothing else. As soon as I dropped her off, I turned the car around and headed straight for Cairns Hospital.

I walked into the emergency department and checked myself in, knowing this time I couldn't afford to wait any longer.

That's where the real suffering began.

I spent eight hours in that waiting room. Eight long hours sitting under harsh fluorescent lights that made everyone look unwell, in chairs that felt designed to punish rather than support. Time stretched in unnatural ways. Minutes crawled past like hours while my body deteriorated and my mind circled the same question over and over again: *what is happening inside me right now?*

There's a particular kind of fear that settles in when you're in an emergency department and you know something is wrong, but no one has told you what yet. You're not dramatic enough to be rushed, not stable enough to relax. You just sit there, waiting, wondering whether whatever is happening inside you is getting worse with every passing minute, quietly doing damage while the world moves around you at a normal pace.

The nurses moved me in and out for tests in a steady rotation. Blood work. IV lines. More blood. Pain medication that barely took the edge off. By then, the pain had returned with force. Whatever the tablets had been masking over the weekend were gone, and whatever was happening inside me was accelerating.

The sensation had shifted from something sharp but tolerable to something overwhelming. By Monday afternoon, it wasn't just pain anymore. Breathing took effort. Sitting upright was exhausting. I could feel my body struggling to keep up, minute by minute, as if it was fighting something it no longer had the strength to contain.

At 1633 they finally wheeled me in for a CT scan. By the time I was back in the room less than half an hour later, the atmosphere had shifted. Nurses moved faster. A doctor returned with another doctor beside her. Voices dropped, sped up, and carried a sense of urgency that hadn't been there before.

Then the door opened and a doctor walked in. She looked to be in her early forties, calm, composed, wearing blue scrubs. Her expression was controlled, professional, and careful. It was the tone medical staff use when they are about to deliver serious news and are trying to keep the room steady while they do it.

"Hi Matthew, my name is Dr Ravi," she said. "Has anyone spoken to you yet?"

I shook my head and told her no.

She pulled a chair up beside the bed and sat down. She wasn't rushed and she wasn't casual either. It was the kind of deliberate calm doctors use when they are about to say something serious and want to make sure you are steady enough to hear it.

"Okay," she said. "I'm going to explain what we've found."

She didn't draw it out. She didn't soften the language.

"My team has reviewed your CT scan," she said. "Your appendix has ruptured."

Ruptured. Three centimetres. Completely burst. Gone.

I don't know exactly why I started crying. It might have been relief. It might have been fear. It might have been two days of pain and confusion finally colliding at once. It wasn't sobbing, and it wasn't hysteria. Just tears.

For the first time in days, someone had given me an answer. I hadn't imagined it. I wasn't losing my mind. Something was wrong, and it was serious.

Almost immediately, that relief was overtaken by fear. A burst appendix isn't something you manage at home or walk off with painkillers. It means infection spilling into your body. It means poison spreading. It means the kind of thing people used to die from before modern surgery made survival routine instead of luck.

In that moment, hearing the words spoken out loud, the thought that landed was simple and unmistakable. This was bad in a way that meant things were about to move fast whether I was ready or not.

"We need to take you to emergency surgery," she said. "Tonight. As soon as the theatre is ready."

At 2000, I was taken into surgery.

When I woke around 0230, I was groggy and disoriented, hovering somewhere between consciousness and pain. The surgical team had done their work. The infection had been cleared, what remained of the appendix removed, and my abdomen stitched back together. Forty staples ran down from my belly button, leaving a scar that looked less like a medical procedure and more like the aftermath of something violent.

Waking up wasn't the end of it.

It was the beginning of a different fight entirely.

I spent the next few days tethered to IV lines while they pumped antibiotics into me around the clock. It should have stabilised things. It didn't. My body rejected the first course violently. Fever spiked. Shaking set in. The pain surged back instead of easing. Whatever they were giving me wasn't working, and my system made that clear fast.

They changed antibiotics and tried again.

That was when the next complication arrived. Post-operative ileus.

Post-operative ileus is the kind of complication you never hear about unless it happens to you. After surgery, your gut can simply stop working. Digestion halts. Nothing moves. Your system shuts itself down without warning or negotiation.

That was how I found myself sliding into one of the most confronting medical experiences of my life.

And no, this wasn't about the catheter.

I handled the catheter the first time like a professional. No drama. In and out. Done. Or so I thought.

Not long after they removed it, my bladder decided it was done cooperating. Completely shut down. I spent the next nine hours trying to piss, with that awful sensation that your back teeth are floating while absolutely nothing happens. No relief. Just mounting pressure and the slow realisation that my body had slammed the brakes on.

Eventually, a nurse looked at me with the kind of expression that tells you the conversation is about to go somewhere unpleasant.

"We'll need to put the catheter back in," she said.

I've dealt with some things in my career, but watching someone lube up a tube and line it up where it should never go while you're already in agony is a very specific form of psychological warfare. There's no bravery in that moment. There's just acceptance.

And somehow, that still wasn't the worst part. That came next.

The nurse told me they needed to insert a nasogastric tube. For anyone lucky enough not to know, that's a tube they push up your nose, down the back of your throat, and into your stomach to drain or feed when your gut has decided to shut down completely.

My gag reflex is instant and aggressive. Eating a banana at the wrong angle has me dry-reaching. So when the nurse calmly told me to "just swallow the tube", I looked at her like she had asked me to swallow a twelve-inch garden hose on purpose.

She started feeding it in, and my body reacted instantly. My eyes watered, my throat said fuck that and every instinct I had screamed no. Before I could warn her, I projectile vomited all over the nurse.

To her credit, she didn't flinch. She didn't swear or step back or even hesitate. She wiped herself off and continued feeding the tube down my throat like this was a completely normal Tuesday afternoon and not a grown man expelling his soul onto her scrubs.

Eventually, after what felt like an hour of pure torture, the tube was finally in place. And once it was in, that was it. It stayed there.

For twelve fucking days.

For twelve days, I lived with a tube down my throat, draining anything that triggered nausea because my gut had completely shut down. Eating became about survival, not recovery. Everything was liquid. Jelly. Custard. Yoghurt. Broth. Food meant for toddlers or people who couldn't chew, not an adult body trying to heal itself.

There was nothing solid, but it wasn't misery either. Ice cream five times a day was objectively excellent. Custard and jelly went from childhood afterthoughts to genuine highlights, and for a while I leaned into that small comfort because there wasn't much else to enjoy.

Still, it wasn't real nourishment in the way you need when you're rebuilding. There was no chewing, no resistance, no sense of being anchored in my body. Everything passed through quickly, soft and sweet and temporary, keeping me alive without ever making me feel whole.

There were needles every day. Antibiotics ran constantly through my veins. Vitals, bloods, infection markers checked on repeat. Nurses waking me at all hours to see whether my body was cooperating or slipping again. I was fighting sepsis and secondary complications while my system tried to reboot after very nearly killing me.

It was relentless, but it wasn't unbroken misery.

Boxing Day landed while I was still in that bed, and somehow I managed to get the Boxing Day Test on the TV. One of the greatest Tests ever played. Then the New Year's Test followed, giving me something familiar to anchor the days. Sessions, overs, breaks. Time measured in something other than medication rounds and blood draws.

Rhiannon brought my laptop in, and for those fourteen days I disappeared into a game of Football Manager. Proper sessions. Long saves. Tactical tinkering that required just enough focus to keep my mind occupied without exhausting me. It sounds small, but it mattered. It gave me control over something, however trivial, when my own body had completely lost it everywhere else.

By the end of the second week, something shifted. The infections started to ease. The pain changed character, becoming less sharp and more distant. My body began to wake up, cautiously, like it didn't quite trust itself yet.

The worst was behind me, but I was still a long way from okay.

Those fourteen days in Cairns Hospital were some of the hardest of my life, physically, mentally, and emotionally. It wasn't just recovering from a burst appendix. It was facing the reality that for the second time in six months; I had come far closer to dying than I was comfortable admitting.

Rhiannon told me later that there was a moment when she genuinely thought she might lose me. Not as a passing fear or a fleeting worry, but in the quiet, heavy way people think when the outcome is no longer abstract. When they start preparing themselves, internally, for a life that looks very different from the one they expected.

That moment stayed with her. And when she told me about it, it stayed with me too.

Because that brush with death wasn't just another medical crisis to recover from. It stripped away the last of my denial. It forced a clarity I hadn't allowed myself before. What followed wasn't a return to normal. It was the beginning of a fundamental change in how I lived, how I led, and how seriously I took the cost of ignoring my own limits.

What came after that was ten months of sustained, consistent effort. There was no dramatic reset, and no moment where everything clicked into place. Recovery unfolded slowly and unevenly, often without any clear sense of progress. Some days felt better. Others didn't. Most of it was quiet, unremarkable work that only made sense once enough time had passed.

As of December 2025, I've lost thirty-five kilograms. At thirty-seven, I'm the fittest I've been as an adult. My thinking is clearer. My mental health is stable. My leadership is stronger.

None of that happened quickly, and none of it happened by accident. It was the cumulative result of decisions made daily, often without motivation and sometimes without confidence they were working at all.

When I walked back into HMAS *Coonawarra* in January 2025, I wasn't lean or shredded, and I didn't look like the kind of transformation people like to post online. What I did have was certainty. I knew that if I kept living the way I had been, ignoring my body, overriding warning signs, and treating health as something separate from duty, I wasn't going to survive the next chapter of my life.

I finally stopped treating health like a personal side project and started recognising it for what it actually is: a leadership responsibility. It isn't something you work on after hours or come back to once the real work is done. It underpins every decision you make and every interaction you have with the people you lead.

Leadership doesn't exist separately from health. It is expressed through it. You can build systems, write frameworks, and memorise principles until you sound sharp in meetings, but none of it holds if your body is failing. None of it survives when you are operating from exhaustion, pain, or a constant threat response instead of clarity and control.

When health is neglected, everything narrows. Patience thins. Judgement loses its edge. You become reactive rather than deliberate, spending your energy containing problems instead of leading people. Decisions turn short term, conversations become transactional, and leadership shrinks into damage control.

When health is prioritised, the opposite unfolds. Stress becomes something you can carry rather than something that overwhelms you. Decisions sharpen. You respond instead of reacting. You show up grounded, consistent, and present, not because you are pushing harder, but because you are no longer fighting your own body just to function.

I wasn't just stepping into a new rank. I was rebuilding myself with intent. Clearer. More disciplined. More present. Determined to lead with integrity and empathy, but also with boundaries, including boundaries with my own body.

That didn't come from motivation or willpower. It came from systems. Small, repeatable choices made daily, long before any visible results showed up. Choices that stabilised my body first, and then my mind. Those systems became the foundation for everything that followed, both in how I lead and how I live. I'll break them down properly later, when I introduce the Operator System, because they didn't just change my circumstances, they changed my trajectory.

What I learned the hard way is this: health is not optional, and it is not something you earn once the work is done. It is the foundation that makes all of the work possible.

Twelve

$226 Well Fucking Spent

Institutions will try to preserve the problem to which they are the solution.
— Clay Shirky

By the end of January, still recovering from surgery, I was back in uniform and back at HMAS *Coonawarra*, stepping into my role as Chief Petty Officer Regulator within H&C.

The posting made sense. The work was familiar, and the expectations were clear. After the chaos of the previous year, it felt like forward movement had finally resumed. There was no sense of relief or closure, just momentum and a renewed focus on rebuilding trust, both with my team and with myself.

I didn't ease back in. I got to work.

Progress came quickly. Outputs were strong and the department stabilised. I supported my people, set direction, aligned standards, and rebuilt our Vision, Mission, and Values. I did the job chiefs are meant to do. I showed up, made decisions, and carried responsibility.

For a brief moment, it felt like the story might end there. A hard year survived, a lesson learned, a return to purpose. The kind of ending you earn, close the cover on, and move forward from.

But the Navy doesn't do neat endings. It doesn't forget. And it rarely lets anyone walk away without settling the account first.

Before I could move on, the system wasn't finished with me. There was still an unresolved matter, a moment it hadn't finished accounting for, one that refused to stay in the past. No matter how much time had passed or how much distance I'd put between myself and that year, one day kept resurfacing.

8 July 2024.

This chapter isn't about punishment for speaking up. That story has already been told, and it doesn't need repeating. What follows is about accountability, not explanation. About the part of what happened that belonged to me, whether I liked it or not, and what it meant to face that honestly instead of pretending it didn't exist.

Toward the end of February, I was called in to speak with the Naval Police Coxswains at HMAS *Coonawarra*. The conversation wasn't hostile, but it wasn't casual either. It was deliberate and measured. The kind of interaction where everyone in the room already understands why they're there, even if no one says it outright.

They asked me to walk them through my version of events. What I understood about insubordination. Where I believed the line sat, and whether I knew I had crossed it.

I told them the truth.

I understood rank. I understood hierarchy. I understood exactly what insubordination was. I'd lived inside that system long enough to know how it worked and why it existed. What I admitted, somewhat dryly, was that I hadn't realised describing someone's leadership style as authoritarian crossed that line. Apparently, calling a dictator a dictator was the issue.

They didn't look surprised.

The system rarely is. It doesn't operate on shock or outrage. It operates on inevitability. Once a moment like that enters the machinery, the outcome is less about discovery and more about process catching up.

Then they handed me the witness statements.

Reading them was confronting, not because they contradicted my memory, but because they exposed how the same moment can fracture into entirely different realities depending on where you're standing when it happens.

Two of the statements aligned closely with what I'd already described. The book slammed on the desk. The line about not caring who was upset. The escalation. The pressure building, minute by minute, until something finally gave. Those accounts captured the context. The tension. The slow accumulation that made the outcome feel inevitable rather than spontaneous.

Three other statements did not align.

In those versions, the context disappeared completely. There was no build-up, no provocation, no environment. I wasn't a man pushed past a limit. I was an enraged hulk. A sudden, volatile presence who erupted without warning.

Details were flattened or removed to make the moment cleaner. Simpler. Easier to condemn.

In three statements, the words *"with all due respect, Ma'am"* were gone entirely. Not paraphrased or softened. Just removed. What remained was the accusation itself, delivered as if I had hurled it across the room like a dagger. No qualifier. No attempt at restraint. No indication that I was still trying, however imperfectly, to operate within the language of rank and respect.

I'm not pretending that phrase gave me a free pass. It didn't. You don't get to say whatever the fuck you want just because you preface it politely. I understand that. I've owned that from the start.

But it mattered that it was said. Because its absence changed the shape of the moment. It turned a controlled but exhausted line into a sudden act of aggression. It recast intent. It stripped away the last signal that I was still trying to speak *up* rather than simply lash *out*.

One statement even claimed I slammed a book onto the table. A book I never carried. I don't bring books into meetings. I use a notepad. I always have. But once something like that is written down, accuracy becomes secondary. The image does the work.

The cumulative effect was clear. The story no longer began with what was said to me, or how long the pressure had been building. It began and ended with my reaction.

Cause and effect collapsed into a single snapshot. Everything that made the moment possible was erased. What remained was a version that was easy to process and easier to discipline: an angry sailor, out of control, acting without justification.

A version that required no one else to answer for anything.

Only one statement didn't fit that mould.

PO Jacinta Gunn's.

She didn't excuse what I did, and she didn't attempt to clean it up or soften it. She acknowledged the moment as it was, including her belief that I was trying to protect her from the onslaught that was coming from the OIC. That detail mattered, not because it justified my behaviour, but because it spoke to intent rather than just impact.

More importantly, she documented what led up to the moment instead of pretending it appeared out of nowhere. She recorded the pressure in the room, the environment that had been building for months, and the tone that existed before anyone raised their voice. Her statement recognised that what happened did not begin with my words, even if it ended with them.

Her account didn't flatten the day into a single outburst or isolate my reaction as a standalone event. It treated what happened as a sequence, where cause existed before effect and context mattered. That alone set it apart from the others.

The contrast between her statement and the rest was impossible to miss. Side by side, they revealed how the same moment can fracture into entirely different versions of reality depending on who is telling the story. And more confronting than that was the realisation that accuracy has very little influence over which version survives.

What survives is the version that requires the least disruption to power.

Sailors protect sailors. Officers protect officers. When those instincts collide, rank determines which version carries weight. That isn't bitterness or hindsight speaking. It's structural reality. The room closes around hierarchy, not context, and once that happens the story stops being about what occurred and starts being about what the system can safely acknowledge without having to look any further up the chain.

That realisation landed. There was no anger in it, no surprise. Just clarity. Once the statements were read side by side, the outcome was no longer uncertain. The process wasn't weighing truth against truth. It was balancing order against disruption, and order always wins.

From that moment on, I understood how this would end. Not because I believed I was blameless, and not because I thought the system was malicious, but because I finally saw the mechanism clearly. The system wasn't asking what happened. It was deciding what it could contain.

Everything that followed was just administration catching up with that decision.

Once I'd read the statements, I knew I was fucked.

I wasn't shocked. I'd known from the moment I walked into that meeting that there would be a cost. What changed after reading the statements wasn't the fact that consequences were coming, but the scale of them. How this would be framed. How far the system would need to go to reassert control.

At that point, the outcome stopped being about truth or fairness. The system had no other way to resolve a moment like that without drawing a hard line. Public challenge to authority, conflicting accounts, rank involved. There was only one lever left to pull.

The question wasn't whether I'd be charged. That was inevitable.

The only unknown was how hard the system was going to make the point.

In March, the paperwork caught up.

I was formally charged under the Defence Force Discipline Act, section 26(1), engaging in insubordinate conduct. The charge was delivered by CPONPC Richford, who was doing exactly what her role required of her. There was no theatre in it. No raised voices. No attempt to moralise the moment. It was simply the system reaching its conclusion and carrying it out.

The wording of the charge itself was narrow and blunt. It stated that I had called a superior officer a dictator, or words to that effect. That was true in the most literal sense. I did use that word. What it didn't capture was how it was said, or why. Context had already been stripped away by that point, and the charge reflected that reduction. Language without sequence. Outcome without lead-up.

By then, there was no practical mechanism to correct that framing. I was advised that challenging the wording would require pleading not guilty, and that choice carried the risk of significantly harsher outcomes. I wasn't interested in gambling my career to argue nuance once the process had already decided what it needed to hear.

So I accepted it as written. Not because it was a complete account of the moment, but because the system no longer had any use for completeness. It had what it needed to proceed.

By that stage, the charge felt less like an accusation and more like an administrative inevitability. The moment had been absorbed, translated, and reduced to a line item that could be actioned. Whatever complexity had existed on the day itself was no longer relevant. The system had moved on, and now it was my turn to meet it where it stood.

Once the charge was laid, I did what any good sailor does when they know they're fucked.

I reached out.

Over the course of my Navy career, I'd built real mateship. Some professional, some far beyond just work. People I'd served with, trained with, led, and been led by. Relationships forged through long days, hard calls, shared pressure, and mutual respect. When I understood what was coming, those were the people I turned to.

I contacted the people whose judgement I trusted. People who had seen me lead when things were steady and when they weren't. People who had watched how I operated under pressure and understood my character outside of a single moment. Some were senior within Defence. Some held influence. Others were civilians who had worked alongside me closely enough to know the kind of leader I actually was, not just the version captured on a charge sheet.

My ties to Navy Football mattered here. Defence sport strips rank away, and that space had allowed me to contribute on equal footing for nearly twenty years. I'd been part of it as a player, a coach, a committee member, and a major sponsor.

I reached out across that network not to ask for favours, but to ask for honesty. I wanted people who would speak to the impact I'd actually had, not dress me up or smooth the edges. I wasn't looking for polish. I was looking for truth.

I cast the net wide.

Senior leaders within Navy Football. Local Northern Territory ministers who had worked with me and seen the way I operated. My coaches from the Entourage, who had watched me rebuild myself. Junior sailors who had served under me and experienced my leadership day to day, not just when things were easy.

I didn't hide from the context. I gave it to them plainly. Most of them already knew what had happened anyway, but I made sure there was no ambiguity about why I was asking. I laid out the charge, the circumstances around it, and the fact that I had crossed a line.

Then I asked them all for the same thing.

Not defence or justification or sympathy.

I asked them to acknowledge the charge as it stood, and then give their honest account of me as a human being and as a sailor. How I showed up. How I led. How I treated people.

I wasn't looking to erase what happened on 8 July. I knew that moment would stand on its own. What mattered was whether that moment was allowed to define my entire career, or whether it was understood in proportion to everything that came before and after it.

Why did I approach so many of my associates?

Because I wanted the room to feel the weight of it.

Part of it was practical. If this process was going to upend my day, my career, and my reputation, then I wanted the full picture sitting in front of the Commanding Officer, not a thin slice of it. And yes, if that meant he had to work a little harder to get through it all, I wasn't going to lose sleep over that.

But more than that, I wanted something unmistakable to land the moment those references were opened. I wanted it to be immediately clear that 8 July did exist in isolation. That this was not a pattern of behaviour, not a man who routinely lost control, not someone who needed correcting as much as containing.

What I was asking the system to look at was proportion. One breaking point set against years of consistent leadership, service, and care for people. A single moment weighed against a record that showed how I managed when it mattered, how I led under pressure, and how I treated those in my charge.

If judgement was going to be passed, then I wanted it passed with the full weight of that context on the table. Not just who I was on one day in July, but who I had been the entire time leading up to it.

I wasn't trying to escape accountability. I was preparing to face it properly.

Four weeks later, I was summoned to appear before the Commanding Officer.

This wasn't a meeting. There was no discussion to be had, no clarification to offer, no version of events left to negotiate or reframe. The process had already decided what this was.

It was a hearing.

I was marched into the room in whites, Chief's cap on, movements automatic after years of repetition. Captain Derek Shivers was standing when I entered. I halted at the mark, saluted, waited for the salute to be returned, and then sat when directed.

The room wasn't hostile, but it wasn't warm either. It was controlled. Procedural. Exact. Emotion had been stripped away so the process could operate cleanly, without distraction. This wasn't about how anyone felt. It was about order, sequence, and authority doing what it exists to do.

Everything personal had already been removed. What remained was structure. And I was now standing inside it.

To my right sat my defending officer, LEUT Allen Dowe. He was one of the best Divisional Officers I've ever served under, not because he bent rules or played favourites, but because he understood both sides of the system. He knew the regulations, the expectations, and the machinery of command, but he also understood the people who had to live inside it. Calm, grounded, and deliberate, he brought exactly the kind of presence you want beside you when the stakes are real.

The prosecution went first.

ABNPC Irine Kellor stood and presented the case. She didn't approach it with hostility or restraint; she approached it with professionalism. Her statement was measured and precise. She addressed the conduct directly, without exaggeration or theatrics, and then placed it in context by acknowledging my record. Nearly two decades of service. Consistent performance. A history of leadership.

It would have been easy to reduce me to the charge alone. It would have been easy to frame the moment as a pattern instead of an exception. She didn't do that. She presented the facts as they stood, without trying to sharpen them into something they weren't.

I'll always respect her for that.

Then Allen stood.

We acknowledged the charge and entered a guilty plea. There was no attempt to dance around it or dilute responsibility. That part was mine to own, and we owned it cleanly.

What followed was Allen doing exactly what a defending officer is meant to do, and then some.

He didn't minimise the charge, but he contextualised it. He walked the Commanding Officer through my diagnosis of Acute Mental Health Disorder. He outlined the environment at FLSE-D and the documented impact it had already had on multiple sailors, long before 8 July ever arrived. He made it clear that this wasn't a man acting out of character for convenience, but someone who had reached a breaking point inside a system that had already been showing cracks.

He then submitted my letter of reflection. Not as an excuse, but as evidence of ownership. Alongside it, he presented his own report of me as my Divisional Officer, grounded in firsthand experience rather than opinion.

And then we moved on to the character references.

They had to be submitted formally, one by one, as evidence.

All forty-six of them.

At that point, I genuinely felt for the poor scribe. Watching them work through that stack, methodically recording each reference, name after name, page after page, I could almost hear the silent resentment forming. But that was the point. There was no rushing past this. No skimming. The weight had to be felt.

Each reference reinforced the same message. This was not a pattern of behaviour. It was a breaking point.

Three of those references came from very senior officers. One came from the Deputy Chief of Navy.

That mattered.

Once the final reference was submitted, the Commanding Officer adjourned the hearing.

There was no commentary, no indication of where it was heading. I was directed to stand, march to the marked position, to face the Commanding Officer, salute, and wait. The salute was returned, and I was marched out of the room.

The adjournment wasn't symbolic. It was practical. The Commanding Officer needed time to read what had just been placed in front of him. Forty-six character references do not skim quickly, and they do not speak quietly.

I waited.

When I was called back in, the atmosphere of the room had changed.

There was no chair waiting for me this time. I was marched to the mark and remained standing at attention. This wasn't a discussion. There would be no clarification, no submissions, no argument.

This was sentencing.

Captain Shivers began by acknowledging my service record. Nearly two decades of consistent performance. Leadership roles held. Results delivered. He spoke directly to the character references, noting both their volume and their consistency. He referenced the seniority of some of the authors, including the Deputy Chief of Navy, and made it clear that this was not something he took lightly.

He stated plainly that my conduct on 8 July sat completely outside my normal behaviour. That the material before him showed a sailor with a strong record, not a pattern of misconduct. That context, history, and character had all been weighed carefully in reaching his decision.

Then he drew the line.

He was clear that regardless of context, publicly challenging a superior officer in that manner could not be allowed to stand. That if behaviour like that was left unaddressed, it risked becoming precedent. And that his responsibility was not only to respond to what had happened, but to ensure it did not happen again, not by me and not by anyone else watching how the system handled it.

He said there were appropriate ways to raise concerns. That the chain of command exists for a reason. And that even when leadership fails, the structure itself must be protected.

Boundaries had to be enforced, and the integrity of the chain of command had to be maintained. That principle, he made clear, sat above individual circumstance.

Then the decision was delivered

A reprimand and a fine of $226.

I acknowledged it, came to attention, saluted, and was marched out. The door didn't slam or echo behind me. It simply closed, and the chapter ended.

The investigation. The charge. The waiting. All of it was done.

What remained wasn't anger or relief. It was clarity; the kind that arrives only once the noise has stopped and the outcome is no longer in question. I didn't appeal the decision or attempt to argue it further. By that point, there was nothing left to debate. The system had done what systems do. It had drawn a line, enforced it, and moved on.

So did I.

When I walked out of that room with the fine notice in my hand, I looked at the number printed at the bottom of the page. $226. Not insignificant but just enough to sting. I paid it without hesitation, not because I agreed with every part of how that day had been reduced, translated, and remembered, but because I understood the trade that had been made.

I had crossed a line. And I had also drawn one.

That fine didn't erase what happened, and it didn't redeem it. It didn't clean the moment up or make it sit neatly inside a narrative anyone could be proud of. What it did do was close the account. The system took its response, I took responsibility for my part, and we both walked away carrying exactly what we were prepared to live with.

$226 well fucking spent.

Thirteen

The Hypothesis

The best way to predict the future is to create it. — Peter Drucker

By the time the fine was paid, the process was complete. The investigation had closed, the charge was resolved, and the system had finished with me. What followed was not relief or anger, but a clear understanding that the machinery had done exactly what it was designed to do.

The story so far does not expose the failure of one leader, one unit, or one isolated decision. It reveals a recurring pattern I had now seen play out across roles, postings, and environments that looked different on the surface but failed in the same way underneath. Leadership was being carried by personality instead of structure. Authority was shaped by temperament rather than clarity. Accountability became emotional instead of procedural. Pressure was absorbed by people rather than being held by systems.

Once I saw that pattern clearly, the question stopped being why this had happened to me. That question had no useful answer. The more productive question was why it kept happening at all.

More importantly, I wanted to understand what would prevent it from repeating.

That is where this chapter begins.

Before going any further, one distinction matters. Traditional military leadership has an essential place. In environments where seconds matter and lives are at risk, such as combat, boarding operations, or emergencies at sea, command and control leadership is not optional. It is necessary. In those moments, clarity must be absolute, authority must be accepted without hesitation, and decisions must be executed immediately. Ambiguity costs lives, and delay compounds risk.

That model exists for a reason, and when it is used in the environments it was designed for, it works.

What I observed over time was not the failure of command and control, but its unexamined use in environments it was never designed for. I saw the same leadership approach applied unchanged in shore postings, training establishments, and office settings where the work was sustained, complex, and human centred. In those environments, success depended less on immediate obedience and more on judgement, consistency, and trust built over time.

When command and control was used there, it did not disappear, but it lost precision. Instead of sharpening thinking, it narrowed it. Instead of building capability, it produced compliance. Pressure flowed downward rather than being absorbed by structure, and the cost of that pressure appeared in people long before it appeared in performance.

The pattern was consistent. Standards did not fail. Discipline did not fail. What failed was adaptability. Leaders were leading every environment as if it were a crisis, and in doing so they quietly turned everyday operations into one long, sustained emergency. The pressure never released. It accumulated.

That gap, between how leadership functions in moments of acute danger and how it needs to function in sustained peacetime operations, was the gap I wanted to test.

By the time I reached the end of that part of my life, I was not looking for motivation, healing, or closure. I had already learned that becoming more resilient inside a broken system simply extends how long the system can continue breaking people. I had paid the price for that lesson physically, mentally, and professionally.

What I wanted instead was something concrete. Something that could be applied deliberately, observed in practice, measured over time, and taught to others.

When I stepped into my next role, I did so with a clear hypothesis formed from experience rather than theory. I believed that servant leadership, when paired with disciplined systems and repeatable frameworks, could outperform traditional command and control leadership in a peacetime military environment. Not because the idea itself was new. I had worked for leaders who instinctively operated this way long before anyone gave it a name.

What was different was the structure.

What I wanted to test was whether those instincts could be made explicit, deliberate, and transferable. A framework that did not rely on personality, goodwill, or individual talent. A system that could be taught, replicated, and sustained even when the leader was tired, under pressure, or eventually gone.

This was about building something clear, deliberate, and structurally sound. It was not theory or belief. It was a test.

By that point in my career, I was not guessing anymore. Time spent leading in both Navy and business had exposed patterns that repeated regardless of environment, industry, or uniform.

The best leaders I worked for in both worlds were operating inside the same underlying system, even if they never named it. They built clarity around intent, set explicit standards, and created environments where accountability was predictable, and performance could compound over time.

The strongest military leaders I served under did more than issue orders. They built structure around the work. They set clear intent and explicit standards. They documented how things were meant to be done and why. Accountability was not personal or emotional; it was tied to agreed expectations. As a result, their teams could function without constant supervision because people understood the mission and their role within it.

Outside the Navy, the most effective business leaders I knew were doing the same thing. They designed environments for consistent performance. Vision was clear, processes were documented, metrics were visible, and the work followed a rhythm that reduced friction. Leadership was repeatable rather than dependent on individual effort.

For a long time, I did not fully understand why this stood out to me. Eventually, I realised it was because I had not grown up inside a single system. My understanding of authority, performance, and people was not shaped by one environment alone. I had enough exposure on both sides to see what each did well and where each quietly fell short.

Once that pattern became visible, I could not unsee it.

The connection between exceptional military leadership and exceptional business leadership was not accidental. It was structural. Defence, from what I have seen, builds process exceptionally well. Where it often falls down is in training those processes properly or consistently following them once pressure arrives. Civilian organisations, particularly strong ones, tend to do Vision, Mission, and Values well, but from my experience, standard operating procedures are often treated as a foreign language rather than a foundation.

What became obvious to me was that each world held half of what worked. The military understood structure, discipline, and process. High-performing civilian organisations understood alignment, clarity of purpose, and human-centred leadership. What was missing was a way to combine the strengths of both into something coherent, practical, and teachable.

The simplest way I can explain what became the Operator System is this. It is the best of both worlds, deliberately combined and organised into a structure that leaders can actually learn, apply, and sustain under pressure.

I was ready to run the experiment.

Once I saw the pattern clearly, I stopped treating it like an insight and started treating it like a design problem. Insights are interesting, but they do not change outcomes on their own. Design does.

I began by documenting what I was already doing, because parts of what would later become the Operator System were already in use. At the start, it was rough and incomplete. I identified five core layers that I knew worked based on my experience.

From there, I deliberately expanded my thinking. I drew on what I had learned through the Entourage, through people like Dan Martell and Alex Hormozi, and through years of reading, observing, and testing what actually worked in real environments. Some ideas stayed. Others were discarded quickly. As gaps became obvious, the structure grew to seven layers, then shifted again as I realised what was missing or redundant.

Even now, as I write, I am aware that no system is ever truly finished. Design evolves as environments change. But this version represents the most stable and effective form I have seen work under pressure. It eventually settled at ten layers, not because ten sounded complete or impressive, but because that was the minimum required to make the system both teachable and resilient when tested in real conditions.

At that point, it stopped being a framework and became an operating system.

It was built for operators, leaders inside the work carrying deadlines, pressure, and people all at once. It was designed to be installed, run consistently, and relied on when conditions were imperfect.

Once established, it made leadership clearer. Decisions accelerated because intent and standards were already defined. Expectations were explicit rather than assumed, which reduced noise and prevented ambiguity from turning into conflict.

I was never under the illusion that a system could do the work on its own. No operating system functions without an operator. You still have to show up. You still have to hold standards. You still have to have the hard conversations and make the decisions others avoid.

The system does not replace leadership.

It supports it.

When the system was fully mapped, it was still theory. It was structured, coherent, and logical, but it had not yet been tested. On paper, it was a ten-layer model and a set of assumptions about how people and teams should operate. Leadership, however, does not live on paper. It lives under pressure, with imperfect people, inside imperfect systems.

That was why it had to be tested. Not gently, and not in a classroom, but in a real department with real constraints, real expectations, and real consequences. If it could not survive everyday military reality, it was not worth keeping.

HMAS *Coonawarra's* H&C department became that testing ground.

No system, no matter how well designed, works in isolation. It needs an environment that allows it to take hold long enough to demonstrate whether it can stabilise, adapt, and perform. That does not mean a perfect environment, but it does mean one that is not already collapsing.

What mattered about what I walked into was that the department functioned. Standards were in place, relationships were intact, and the work was getting done. People were tired, but they were not broken. That foundation allowed the system to be tested honestly, not as a rescue tool, but as an operating model. The question was not whether it could save a failing team, but whether it could take a capable one and make it consistently high-performing without grinding people down in the process. That was the test that mattered.

And it was one I was finally in a position to run.

That was because of CPOML-C Matty Cole.

This was the second time I had taken over from Matty. I had followed him into Food Services in 2021, and once again I inherited a team that was stable, disciplined, and capable. That doesn't happen by chance. It comes from leaders who do the unglamorous work, hold standards consistently, and protect culture without needing to perform it. Matty did exactly that.

That foundation mattered, not because systems require perfection, but because sequence matters. The Operator System can turn fractured teams around. I have no doubt it could have rebuilt the 2024 FLSE-D environment into a high-performing one, but inheriting a team like this meant I could test the system for optimisation rather than recovery.

What Matty left behind meant I didn't have to start with containment. I could start with optimisation. The system could be tested for performance rather than recovery, which allowed its strengths and weaknesses to surface faster and more honestly.

I've always lived by a simple rule in uniform: leave it better than you found it. Applied consistently, that approach compounds over time. Defence has been a good place for me to work, and I've valued the vast majority of my career. In my experience, the difference between a good organisation and a genuinely great one often comes down to small, deliberate shifts in how leaders think about people, systems, and responsibility.

Matty lived that principle. I walked into the evidence of it.

While writing this book, and just before my discharge, he sent me a message; the following sentence stood out to me. Paraphrased, it said, *"Chef, I've had the pleasure of leaving two postings in your capable hands and watching both grow and operate more proficiently than I ever had them going."*

Coming from someone who understood the work and the weight of leadership as deeply as he did, that mattered. Not as validation, but as confirmation that the foundation was real.

And that was exactly what made this the right place to test what came next.

I already knew most of the chefs when I stepped into H&C. I'd spent time with the department after FLSE-D, so this wasn't a cold start. I understood the people, the rhythm of the work, and the operational reality well enough to know what I was stepping into. What I didn't yet know was how far the team could go, or how open they would be to operating in a way that felt unfamiliar inside Defence.

What I was about to introduce wasn't how leadership is usually expressed in uniform. It was structured, explicit, and systemised, closer to how high-performing businesses operate than traditional military practice.

The conditions were right. I had inherited a strong foundation, I had a system ready to test, and I had rebuilt enough physical and mental capacity to execute properly rather than just endure. So instead of hedging, I committed fully.

I gave myself twelve months.

I pushed my discharge date to 19 January 2026 and treated the next year as a live experiment. Not a pilot. Not a trial run. A full commitment to seeing whether this approach would actually hold under real conditions, with real people, inside a real military unit.

Those twelve months were about answering a specific set of questions. Could the Operator System actually work as designed? Could servant leadership, when paired with disciplined systems, outperform command-and-control in a peacetime environment? Could the invisible structures I had learned to recognise be made visible, repeatable, and teachable without relying on personality or goodwill?

I put it to work.

The Operator System fucking dominated.

The H&C team exceeded every performance benchmark I had expected, both operationally and culturally. The results were not anecdotal or self-assessed. In September 2025, I invited the Director of Navy Culture to conduct the same survey that had previously been used at FLSE-D. The contrast between the two sets of results was stark. The data confirmed what we were living day to day. Engagement was higher. Trust was stronger. Clarity was measurable. Performance followed.

Over those twelve months at HMAS *Coonawarra*, the Operator System was not only tested, it was also refined. Weak points were exposed and corrected. Assumptions were pressure tested. What worked was kept. What did not was removed. By the end of that period, this was no longer a concept or a hypothesis. It was a functioning operating system with proven results.

What follows in the rest of this book is everything I implemented, exactly as I implemented it. Not theory or inspiration. The actual structure, decisions, and systems that turned a capable team into an exceptional one.

Now, I will show you exactly what that system is.

Fourteen

The Operator

Out of suffering have emerged the strongest souls; the most massive characters are seared with scars. — Khalil Gibran

Out of my own suffering came the birth of the Operator System.

It was not designed in isolation or built in theory. It was developed while leading H&C, refined under sustained pressure, and used daily in a real operational environment. I relied on it to make decisions quickly, to create clarity where it mattered, and to reduce the load that normally lands on a single leader.

The most important outcome was not speed or efficiency. It was sustainability. When the system is built properly, the leader stops being the point of failure. The team reaches a level of clarity and capability where work continues without constant intervention. Instead of operating inside the department, I was able to operate on it.

As the system matured, I deliberately stepped out of the middle. The system carried the weight instead of me. The team did not stall. It performed.

What follows is not a concept or a mindset. It is the structure I built, used, and trusted. This chapter shows the system as a whole, before each part is broken down and taught in detail.

To understand how the system works, it helps to first understand how I define an operator.

What is an Operator?

In the military, an operator is not someone who simply follows orders. An operator is trusted to act when conditions are imperfect, information is incomplete, and consequences are real. They are expected to deliver outcomes with limited resources, compressed time, and minimal supervision.

Execution matters, but it is not what separates strong operators from average ones. The difference is the ability to move deliberately between leading people and managing work, without defaulting to one out of habit.

I saw that balance early in my career with leaders like Lieutenant Commander Brigg. He looked after his people and delivered results at the same time. Mission and people were not competing priorities. The work got done, and the team held together while it was happening.

The System itself

That requirement forced me to build a system.

A ten-level operating model designed to make leadership runnable day to day, not just discussable when conditions are calm. Each level builds on the last, because performance without sequence does not last.

At every level, the system answers the same two questions: how I should operate here, and what must be in place so the team can function without constant intervention.

OPERATOR SYSTEM

THE MASTERY
Lead like a pro

THE EXECUTION
Build the rhythm

Level 9
Daily Systems = Self Management

Level 8
Battle Rhythms = Management

Level 7
One-On-One Meetings = Management

THE STRUCTURE
Build the how

Level 6
Individual Goal Setting = Leadership

Level 5
Department Goals = Management

Level 4
Accountability Charts = Leadership

THE FOUNDATION
Build the why

Level 3
Processes & Procedures = Management

Level 2
Standards & Expectations = Leadership

Level 1
Vision, Mission & Values = Leadership

THE FOUNDATION (Levels 1–3): Build the Why

When I finally laid the full system out, one thing was immediately clear. Everything inside the Operator System depends on what it is built on. If the foundation is done properly, a department, unit, or business does not need the leader sitting in the middle of every decision. The system holds because the base is strong enough to carry the weight.

That understanding did not come from books or theory. It came from years spent inside departments where no one could clearly articulate what they were actually working toward. We usually understood the broader mission. We all knew the same inherited corporate values. What was often not clear was the leader's vision, the team's vision, or how any of it translated into day-to-day decisions.

When pressure increased, there was nothing consistent to return to. Decisions became personal rather than principled. Standards shifted depending on who was present. People hesitated or guessed because the ground beneath them had never been clearly defined.

That experience shaped the foundation of the Operator System.

Vision, Mission, and Values had to sit at the base, not as slogans or laminated statements, but as something practical enough to guide decisions when conditions were uncomfortable. That foundation then needed reinforcement through clear Standards and Expectations, and support through documented Processes and Procedures.

This layer is about reliability.

What Are We Standing on?

When the foundation is built properly, people know why they are there, what is expected of them, and how the work gets done. Once that is established, everything above it becomes possible.

The foundation is made up of three elements, built in this order.

Level 1: Vision, Mission and Values

This defines why the team exists in terms people can genuinely align to. Not slogans, but shared meaning that guides decisions when conditions are difficult.

Level 2: Standards and Expectations

This is where values become behaviour. Standards make expectations visible and define what good looks like, especially when pressure is high.

Level 3: Processes and Procedures

This is where work becomes repeatable. Clear processes reduce reliance on memory or heroics and allow performance to hold regardless of who is on shift.

When these three elements are in place, people stop guessing. They understand why they are there, what is expected of them, and how the work gets done, even when the leader is not present. Everything else in the Operator System is built on this layer. Without it, nothing above it lasts.

THE STRUCTURE (Levels 4–6): Build the How

Once the foundation is in place, the nature of the system changes. The work is no longer about purpose or standards, but about movement. Effort needs direction, responsibility needs definition, and work needs a shape that allows it to move forward without constant intervention.

This is the point where I consistently saw individuals stall, even in teams that cared deeply about what they were doing. Motivation was rarely the issue. In my experience, people do not come to work asking how they can undermine the place. Most want to contribute and do good work. What they often lack is a clear frame that shows how their individual effort connects to departmental, organisational, or business outcomes.

Without structure, accountability tends to drift toward the highest performers. Ownership becomes implied rather than explicit. Decisions slow because authority is unclear. Some tasks overlap while others are quietly missed. Everyone stays busy, but coordinated progress becomes difficult to measure or sustain.

I learned that this was not a people or a commitment problem. It was a structural problem.

This layer of the Operator System exists to give work a clear shape. Not through added bureaucracy, but through deliberate clarity. When structure is done properly, people know what they own, what they are working toward, and how their role fits into the wider system. Effort starts to align rather than collide.

This is the point where leadership stops relying on hope and starts relying on design. Work moves because the structure supports it, not because someone is constantly pushing from the middle.

The structure layer is built through three elements.

Level 4: Accountability Charts

This is where ownership becomes explicit. Every function, responsibility, and outcome has a clearly assigned owner. Ambiguity is removed. When accountability is visible, friction drops and execution improves because people no longer need to defend territory or second-guess responsibility.

Level 5: Department Goals

Once ownership is clear, direction matters. This level introduces defined ninety-day goals that translate purpose into priorities. These goals are short enough to remain grounded in reality and long enough to create momentum. They give the team a shared focus and a measurable way to assess progress.

Level 6: Individual Goal Setting

This is where structure becomes personal through the leader's lens. The role of the manager here is to deliberately link individual goals back to department, organisational, or business objectives. Performance and development are not separate conversations. Growth is built into the work itself, and each person can see how their effort contributes to outcomes that matter.

When these three elements are in place, alignment begins to emerge naturally. Work, people, and direction move together without constant correction. Ownership is clear. Goals are visible. Development is intentional.

At that point, the team stops operating reactively. The system is doing what it was designed to do.

Only then is it possible to introduce rhythm.

THE EXECUTION (Levels 7–9): Build the Rhythm

With a stable foundation and clear structure, execution takes shape through rhythm. Work follows a predictable pattern, effort moves with intent, and the team operates inside a shared cadence rather than reacting day to day. Structure stops living on paper and starts showing up in how the work is done.

When I introduced this layer, the effect was immediate. We began operating consistently. Issues surfaced earlier with less emotional weight. Conversations became clearer and more direct. Performance discussions happened in their proper place rather than being carried informally or delayed. People knew where and when work-related conversations belonged.

That is what rhythm provides. It gives work somewhere to land.

Level 7: One-on-One Meetings

One-on-one meetings are the primary interface between leadership and performance. Used properly, they create a consistent space where trust is built, issues surface early, and development happens alongside the work. When these conversations follow a predictable rhythm, feedback becomes easier to receive and act on, expectations stay clear, and performance discussions become a normal part of how the team operates rather than something avoided or delayed.

Level 8: Battle Rhythms

The battle rhythm is the calendar that holds the system together. It embeds alignment into time by defining what the team does daily, weekly, and monthly, and when decisions and conversations belong. With a clear cadence, issues surface in the right forums, priorities are reviewed on schedule, and momentum builds through consistency. As the rhythm settles, work flows with less interruption, decision-making aligns with timing rather than urgency, and leadership operates with foresight because the system carries the tempo.

Level 9: Daily Systems

This level focuses on how the leader operates personally inside the system. Daily systems combine leadership infrastructure with personal productivity, shaping routines, habits, and decision filters that direct attention and energy. With a consistent daily structure in place, critical work is handled deliberately, decisions are made with intent, and priorities are carried by the system rather than the leader's head. As a result, execution no longer depends on constant availability, and the team continues to operate with clarity and stability.

This is when the system becomes fully operational.

MASTERY (Level 10)

Level 10: Lead Like a Pro

I'm not Simon Sinek, and I'm not going to pretend to teach servant leadership as a philosophy. What I can speak to is how people-first leadership brought the Operator System together in practice. Through using ideas I'd absorbed from Sinek's work and others, I found that servant leadership, applied deliberately, was the most effective way to lead once the system itself was in place.

In my experience, servant leadership is what determines whether a system feels imposed or owned. Authority still exists, but it is reinforced by clarity and consistency rather than force. The leader remains accountable but no longer sits in the middle of every decision. Decisions stop stacking at the top, capability grows across the team, and people develop because the environment supports them. That is what mastery looks like in practice. Not control over outcomes, but the ability to create conditions where outcomes are produced consistently by others. This only works because everything underneath it is already in place, and it is why Level 10 is where leadership stops being carried by pressure and starts being multiplied through people.

A Note Before We Go Further

Up to this point, you have seen the full shape of the Operator System. This chapter is an orientation.

What follows is where the work becomes specific. Each level will be broken down in detail, including how it is built, how it is implemented, where it fails when done poorly, and what it looks like when it is done well.

This system does not ask you to change who you are. It gives you a structure to operate with intent, replacing reaction with clarity, and pressure with rhythm.

If you have read this far, you already know whether that is how you want to lead.

Now we build it.

OPERATOR SYSTEM

THE MASTERY
Lead like a pro

THE EXECUTION
Build the rhythm

Level 9
Daily Systems = Self Management

Level 8
Battle Rhythms = Management

Level 7
One-On-One Meetings = Management

THE STRUCTURE
Build the how

Level 6
Individual Goal Setting = Leadership

Level 5
Department Goals = Management

Level 4
Accountability Charts = Leadership

THE FOUNDATION
Build the why

Level 3
Processes & Procedures = Management

Level 2
Standards & Expectations = Leadership

Level 1
Vision, Mission & Values = Leadership

Fifteen

Vision, Mission, and Values — Your Leadership Flag

When words don't guide behaviour under pressure, they're not values. They're decoration.   — Unknown

One week into the role, I stood in front of a whiteboard facing twenty-five sailors and I could see the scepticism in their eyes. They had seen leaders arrive with big energy and bigger promises, only for nothing to change where it mattered.

I did not blame them. I had been in their position, listening to the same style of brief delivered with the same confidence. The OIC had opened her tenure with the usual language about turning things around and winning together. It sounded right in the moment, and we all know how that ended. So I did not ask them to believe in me. I did not try to motivate them into trust.

I told them the truth. I was not there to give another speech. I was there to implement a system, and the first move in that system was to plant a clear leadership flag that matched the work we actually did.

I had learned that most culture issues at department level shrink fast when the team has a Vision, Mission, and Values that people can remember, repeat, and use in real decisions. If the words are vague, people fill the gaps with assumptions. If the words are clear, the team has a reference point.

The Navy's intent is clear enough at the strategic level, but at the deckplate level it often turns into activity without a shared direction. We used to have a mission you could carry in your head, "To fight and win at sea." The current wording might be accurate, but it is not memorable, and if your people cannot repeat the mission, it will not guide how they work.

Before I asked this team to buy into anything, I took my key people off-site and we built our own leadership flag. We defined where we were going, how we would operate, and what behaviour we would defend, in language that made sense for H&C and could survive a normal Tuesday under pressure.

That meant addressing something everyone in the room already felt but rarely said out loud. Defence places a lot of weight on generic, 1970s one-word corporate values that look fine on a poster but do fuck all to guide behaviour on the floor.

They work at the highest, corporate level of the organisation. As a broad statement of intent, they give Defence a shared language and a set of principles senior leadership can point to. The problem is that they only work if people are committed to actively translating them, and that translation often does not happen at the department level.

In practice, those values are too distant from the work to guide behaviour when it actually matters. They don't help a sailor decide how to act in a crowded galley, during a staffing shortfall, or in the middle of a conflict between teammates. When pressure arrives, people don't reach for abstract principles written for the entire organisation. They fall back on habit, personality, or whoever happens to be in charge that day.

My honest view, and it probably means fuck all to the Chief of Navy, is that Defence should never rely on a single set of values to carry an organisation so diverse. A submarine crew, a galley team, a training establishment, and a headquarters staff all do fundamentally different work under different pressures. Expecting one generic set of words to guide behaviour everywhere is unrealistic. If values are going to matter, every department and unit needs its own Vision, Mission, and Values that reflect the reality of the job they are doing.

The same pattern shows up in civilian organisations and businesses. In my experience, a single set of values can work when an organisation is small enough for everyone to feel the same pressures and do broadly similar work. Once you move beyond roughly a hundred people, and especially once distinct departments or sub-teams emerge, that assumption breaks down. Sales, operations, finance, and frontline delivery may share an overarching purpose, but they operate under very different constraints and incentives. Expecting one generic set of values to guide behaviour equally well across all of them rarely works.

What works instead is a layered approach. An organisation can hold a small set of high-level values, but each department needs its own Vision, Mission, and Values that translate those ideas into something operational. When teams do fundamentally different work, values only become useful when they are close enough to the work to guide decisions in real time.

These are the Defence values we are all familiar with:

Service: The selflessness of character to place the security and interests of our nation and its people ahead of my own.

Courage: The strength of character to say and do the right thing, always, especially in the face of adversity.

Respect: The humanity of character to value others and treat them with dignity.

Integrity: The consistency of character to align my thoughts, words and actions to do what is right.

Excellence: The willingness of character to strive each day to be the best I can be, both professionally and personally.

On paper, those values are hard to argue with. They are well intentioned and technically sound. The issue is not what they say, but how they are actually used. Most sailors cannot recall the full definitions, so the values collapse into single words. Once that happens, meaning becomes subjective and behaviour is left to personal interpretation.

A single word like *Service* can mean entirely different things depending on the role. To a warfighter, service may mean duty to country or mission above all else. To me, leading H&C, service meant delivering for our customers every single day to the highest possible standard. Neither interpretation is wrong, but without clarity, people fill the gap themselves.

That is the real weakness. The values are so broad they can apply to almost any organisation, in almost any industry, without changing a word. When values operate at that level, they stop guiding behaviour and start functioning as background noise. They describe everything, which means they direct nothing.

That gap between words and behaviour is the reason I flipped the process entirely.

Rather than starting with abstract language and hoping it shaped behaviour later, I started with what was already happening on the floor. Before formally stepping into the role, I worked with my key people to examine what was genuinely working in the department, not what sounded impressive on paper.

The approach was straightforward. We identified the sailors who consistently delivered results and strengthened the team around them, then traced the patterns behind their effectiveness. How they approached their work. How they interacted with others. How they made decisions when conditions were uncertain or constrained.

We did not invent values. We identified and named behaviours that were already present and already working.

What emerged from that process was something the team immediately recognised. It reflected how their best people already operated and gave clear language to behaviour they could repeat, reinforce, and rely on in daily work.

The Three Pillars Defined

I didn't approach Vision, Mission, and Values as abstract leadership concepts. I approached them as tools that either worked in the environment I was responsible for, or they didn't.

What mattered to me was not whether the words sounded right, but whether they helped people make better decisions during the day, especially when I wasn't standing next to them. If they couldn't do that, they weren't useful.

Over time, this is how those three pillars came to mean something concrete in my work.

Vision defines the change the team exists to create. Not the task, not the output, but the impact. It gives people a reason to care about the work beyond completing another shift. When decisions compete and trade-offs appear, Vision provides direction without needing instruction.

Mission defines how that Vision is pursued day to day. It sets the operating mandate for the team. If Vision explains why the work matters, Mission explains what we do consistently to move toward it. It turns intent into action.

Values define behaviour. Not aspirational traits, but observable actions. They describe how people are expected to operate with each other and with those they serve. More importantly, they define what behaviour will be supported, reinforced, and corrected. If a value cannot be applied to real situations, it has no place in the system.

Those definitions were shaped by the work itself.

H&C was not just about producing meals. It was one of the few constants in sailors' lives during long and demanding postings. It was where people gathered, decompressed, and felt looked after. The department influenced morale far more than most people realised.

If our Vision, Mission, and Values were going to matter, they had to come directly from the reality of the work. Not borrowed language, not borrowed intent, and not something that needed explanation every time pressure was applied.

We built a leadership flag that reflected how the department actually operated, what the work demanded, and how we expected people to show up for each other and for the sailors we served. This was not an exercise in wordsmithing. It was about defining a reference point that decisions, behaviour, and accountability could be anchored to.

Once it was set, everything else aligned around it.

This is what we committed to.

Vision: Feeding the Navy's spirit by crafting the finest hospitality experience.

Mission: Deliver the finest hospitality and culinary expertise by fostering creativity and freedom for our people, always.

Values:

• Adapt & Overcome — Conquer the unthinkable

• Pursue Greatness — Surpass every customer expectation

• United We Achieve — Succeeding as one team

• Serve with Heart — Embody respect, compassion, and humility

• Bring Fun — Find happiness in every moment

• Push the Limits — Embrace innovation

• Evolve & Flourish — Pursue excellence in every endeavour

These were not aspirational statements. They defined how the department would operate, what behaviour would be supported, and what we were prepared to protect.

That was the point where the system stopped being theoretical and started being tested.

VISION · MISSION · VALUES

VISION

Feeding the Navy's spirit by crafting the finest hospitality experience.

MISSION

Deliver the finest hospitality and culinary expertise by fostering creativity and freedom for our people, always.

VALUES

1. ADAPT & OVERCOME
Conquer the unthinkable

2. PURSUE GREATNESS
Surpass every customer expectation

3. UNITED WE ACHIEVE
Succeeding as one team

4. SERVE WITH HEART
Embody respect, compassion, and humility

5. BRING FUN
Find happiness in every moment

6. PUSH THE LIMITS
Embrace innovation

7. EVOLVE & FLOURISH
Pursue excellence in every endeavour

Why This Vision Actually Matters

The Vision we landed on was simple on the surface: *Feeding the Navy's spirit by crafting the finest hospitality experience.*

When we first put it in front of the team, the reaction was predictable. Polite nods. Folded arms. The look that says, *here we go again.* To them, it sounded like more corporate language dressed up as meaning.

I asked a simple question.

"What do you think your job is?"

The answers came back exactly as expected.

"Serve food." "Cook meals." "Run the galley."

All technically correct. All incomplete.

I pushed a little further.

"So that's it? You're just a meal delivery service?"

That's when one of the junior sailors spoke up.

"No, Chief. When someone's having a shit day, sometimes the galley is the only place they feel looked after. We're not just feeding bodies. We're feeding morale."

That is what *feeding the Navy's spirit* actually means. It is not about plates of food. It is about how people feel when they walk through the door. It is the difference between feeling like a number and feeling valued. Between getting through a day and being reminded that someone still gives a shit.

Crafting the finest hospitality experience was never about being elaborate or performative. It was about treating every meal and every interaction as a signal of care. The quality of the food, the presentation on the plate, the tone of service, and the atmosphere in the space all communicated the same message. They told sailors whether they mattered.

When that understanding settled in, the way the team saw their work changed. They were no longer just preparing meals or running a galley. They understood their role as supporting morale, providing consistency, and offering a place where people felt looked after during demanding periods of service.

The Mission: How We Actually Do It

A Vision on its own does not change behaviour. It describes where you are going, but it does not explain how the work is meant to happen once the day starts. The Mission exists to answer that gap. It translates intent into daily action and makes clear what the team is expected to do, consistently, to move the Vision forward.

Our Mission was simple and deliberate:

Deliver the finest hospitality and culinary expertise by fostering creativity and freedom for our people, always.

I was aware that language like creativity and freedom would sit uncomfortably with some people in a military environment. Those words are often treated as risks rather than tools. My experience, both in uniform and outside it, had shown me something different. Excellence does not come from controlling capable professionals. It comes from setting clear standards and then giving people the space to meet them in a way that fits their skill and judgement.

I made that position clear to the team early.

I told them that I was not interested in blind compliance with a recipe book. I cared about the outcome. If someone could improve a dish, adapt a process, or lift the standard of service, I expected them to do it. If they saw a problem, I wanted them to fix it. Initiative was not something to wait for permission on.

If that expectation was going to mean anything, it had to be backed by action. So I removed the menu.

Completely.

The menu we had inherited was not guidance. It was a constraint. It dictated what to cook, how to cook it, and when to serve it, leaving little room for judgement or skill. It made creativity impossible while pretending to support consistency.

In its place, I introduced a protein-based menu. The team was given the available proteins for the week and clear boundaries around nutrition, cost, and timing. Everything else was theirs to decide. How it was prepared, what sides were served, and how it was presented became professional choices rather than instructions to follow.

That shift changed where responsibility sat. The standard did not drop. It became visible. The expectation was simple and explicit: deliver quality and show what you are capable of.

The final line of the Mission, "for our people always", mattered just as much as the rest. This was not only about feeding the base. It was about creating an environment where the H&C team could operate with trust, autonomy, and pride in their work. If the people doing the job were constrained, disengaged, or second-guessed, the quality of what they delivered would always reflect that.

Over time, the Mission became our daily operating mandate. Deliver excellence by empowering people. It guided decisions, shaped behaviour, and removed the need for constant direction.

The results did not show up as one-off events or staged moments. What you see in the photo is not a dining-in night or a VIP function. It is a normal service, delivered to standard, without me standing in the middle of it.

That is what a Mission is meant to do.

The Mission gave the team freedom and direction, but it also created a requirement. If people were going to be trusted with autonomy, there had to be a clear standard for how that freedom was used. That is where values stopped being abstract and became non-negotiable. Values were not there to inspire or motivate. They existed to define acceptable behaviour when judgement was required and supervision was not. They were the guardrails that ensured creativity, initiative, and ownership never came at the expense of respect, trust, or cohesion. Until values are tested in real situations, they are just words.

The first real test came sooner than I expected.

When Values Collide: The Integration Test

About three weeks after we introduced the values, I walked into the galley and noticed something that didn't sit right. The space was busy and loud, but the room wasn't moving as one. A group of sailors were relaxed and joking. A few of the junior sailors were quieter than usual, sticking close to their stations and staying out of the centre of the room.

Nothing obvious was wrong, but the difference was clear enough to pay attention to.

Later that day, I asked one of them what was happening. A couple of the senior staff had been leaning into humour at a newer sailor's expense. Not as a single incident, but as a pattern. Jokes about mistakes. Stories retold for laughs. Comments that landed one way for the people making them and another for the person receiving them. The sort of behaviour that has been normalised for a long time and usually gets brushed off as banter.

I brought the team together and asked one question. Which of our values are we living right now?

One of the senior sailors answered straight away. Bring Fun, Chief. We're keeping morale up and keeping things light.

I let that land, then asked a second question. Where does Serve With Heart fit in this? Are respect, compassion, and humility present at the same time?

The room went quiet. People thought about it.

I laid it out plainly. Our values don't operate in isolation. They only work as a set. You don't get to apply one and suspend the others. If behaviour can't stand across all of them, it doesn't fit. Humour that relies on someone else carrying the weight isn't aligned with how we agreed to operate.

After a moment, the senior sailor spoke up. He said he hadn't looked at it that way and said he owed Woody an apology. He followed through without being prompted.

From that point on, the standard was clear. The values weren't something to reference selectively. They were the framework the team used to check behaviour as it happened. Decisions became easier to make, conversations became more direct, and correction stopped being personal.

From there, the values stopped needing explanation. They were simply used.

'Adapt & Overcome' in Action

Adapt & Overcome sat at the top of our values because it reflected the reality of the work. Hospitality is dynamic by nature. Equipment fails, numbers change, and service still has to be delivered. The question was never whether disruption would occur. It was how the team would respond when it did.

About six weeks into operating this way, that reality showed up during a Tuesday lunch service for roughly two hundred people. Midway through preparation, the main oven failed completely. There was no warning and no immediate fix.

Chefs have always adapted. That has never been the issue. What usually happened in moments like this was that the problem migrated upward until it became the Chief Chef's problem to solve. The team waited for direction, not because they lacked capability, but because ownership lived at the top.

This time, ownership stayed where the work was.

One of the chefs, Ozzie, came to me with a plan already in motion. The oven was down. Maintenance had been called. The menu had been adjusted from roast to pan-seared options using the stovetops. Sides had been matched accordingly. The team had been briefed. Service would run on time.

I asked if he was confident. He was. I told him to execute.

Lunch service ran as scheduled. The food was clean, well presented, and consistent. Feedback that day was stronger than usual. The flexibility built into the mission gave the team permission to adjust the menu without seeking approval, and they used it well.

What mattered was not that a problem was solved. It was how it was handled. Ozzie assessed the situation, made decisions within the boundaries of the system, and moved the work forward with his team. The value did not need to be referenced in the moment because it had already shaped the response.

I did not step into the workflow. I did not direct the adjustment. The system carried the load.

Pavlova for the Dining-in Night created by Chief Chef Jacques

Bringing Your Vision, Mission, and Values to Life

Once the Vision, Mission, and Values were set, the focus shifted to how they would be used. I treated them as working inputs, not background context. If they were going to matter, they had to show up inside the operating rhythm of the department.

The first place they were embedded was the weekly management meeting. From the beginning, values were the opening reference point. Before discussing operations, performance, or emerging issues, we anchored the conversation in how we were operating as a team.

Each leader came prepared to do one of three things: describe how they had applied a value during the previous week, recognise someone else who had demonstrated one, or point to a decision or outcome shaped by a value. This established a shared expectation that values were part of how work was assessed and discussed.

That approach extended beyond the management group. In department-wide training sessions, participation was expected from everyone. The aim was shared clarity around how values appeared in daily work, using real examples drawn from the environment people were operating in.

As the language became familiar, behaviour became easier to recognise and discuss. Decisions, interactions, and outcomes were referenced against a common frame rather than personal interpretation. This reduced ambiguity and supported consistency across the department.

To support this, I documented the values in a format people could use. I produced a Vision, Mission, Values magazine that explained each value in practical terms, showed how it applied in our context, and outlined why it mattered in our environment. I funded the printing myself to ensure it was treated as working material rather than promotional content.

The values were also made visible in the physical workspace. They were displayed in each galley and in my office as a constant reference point. Anyone entering the space could immediately see how the department operated.

Communication channels were then aligned to reinforce that visibility. I set up three Signal groups, an application approved for Defence use, each with a clear purpose. One group was used for announcements and critical information. One, called the Wins chat, recognised progress, effort, and milestones. The third, the Photo chat, was used to share work output, particularly food presentation.

The Photo chat had a direct effect on standards. As chefs shared their work, expectations rose through peer reference. Quality improved through comparison and pride in output rather than instruction. The system created its own feedback loop.

Over time, this led to the creation of the H&C Magazine, which showcased the department's work more broadly. This reinforced standards internally and established a clear external identity. The department became defined by how it operated, not just by its function.

From there, Vision, Mission, and Values were embedded into onboarding, training material, performance conversations, and daily communication. New joiners received the Vision, Mission, and Values material before arrival, giving them clarity on expectations from the outset.

Throughout this process, values were referenced consistently in decisions, recognition, and correction. They became the shared filter through which behaviour and performance were assessed across the department.

How I Built Our Vision, Mission, and Values

If you want the kind of clarity and ownership described above, the work does not start with writing statements. It starts with how those statements are built. What follows is not theory or a framework lifted from somewhere else. It is the exact process I used at H&C to build our Vision, Mission, and Values in a way the team recognised and carried forward.

I approached this deliberately. I did not write Vision, Mission, and Values in isolation and present them as a finished product. Instead, I built them with the people who lived the work every day, using their behaviour, decisions, and standards as the reference point. The context will differ across organisations, but the mechanics of the process remain consistent.

Step one: I chose the right people

I selected a small group of my key people and took them off-site. I chose them because they were the ones the department leaned on when equipment failed, numbers changed, or a service went sideways. They understood the work, they understood the pace, and they felt the cost when standards slipped.

Before we started, I set the intent in plain terms. We were there to capture how we already operated at our best and turn that into a clear reference point for decisions and behaviour. I wanted language that matched what happened on the floor, because anything that didn't match would die the first time we hit a hard week.

Step two: I watched performance before I wrote anything down

In the weeks leading into that session, I spent time watching the strongest performers closely. I paid attention to how they made calls when the plan changed, how they handled pressure during service, and how they treated the people around them when tempo lifted.

I was looking for patterns I could name. When I could see those patterns clearly, the values stopped being something we had to invent. They became something we could document.

Step three: I ran the session through practical questions

When we sat down together, I kept us away from wordsmithing. We worked through the job as it actually existed.

I asked what impact the department had on the base when we were operating well. I asked what behaviours made service smoother when we were short-staffed or dealing with failures. I asked what the best people did that lifted everyone else, and what conditions helped people stay engaged and proud of their work.

Those answers gave us the raw material. We pulled language from real examples, because that kept the conversation honest and kept the output usable.

Step four: I tested everything against our operating reality

As the values started to take shape, I ran each one against the environment we worked in. H&C is fast, public, and unforgiving. People notice immediately when standards slip, and we don't get the luxury of pausing service to regroup.

If a value didn't fit that reality, I didn't keep it. Adapt & Overcome stayed because it matched what happens in a galley. Equipment breaks, numbers shift, and service still goes out. That value already existed in the way our best people operated, so we named it and made it visible.

Step five: I set my own enforcement threshold

Once we had a draft set, I pressure-tested it against my own willingness to enforce it. I asked myself a simple question for each value: if someone kept breaking this, would I be prepared to document it, counsel it, and run the formal process if it came to that?

If the answer was no, I removed it. I wasn't interested in values that sounded good but couldn't survive contact with real accountability.

Step six: I made it visible and used it daily

Once the language was set, I formalised it. I documented it, put it where people could see it, and referenced it in everyday conversations.

I used it in recognition when someone lifted the standard. I used it in correction when behaviour drifted. I used it in performance conversations, so expectations stayed consistent across the department.

After a few weeks, the language showed up in conversations without prompting. It was referenced during service, in one-on-ones, and when behaviour needed checking. I didn't have to introduce it anymore. It was already there.

OPERATOR SYSTEM

THE MASTERY
Lead like a pro

THE EXECUTION
Build the rhythm

Level 9
Daily Systems = Self Management

Level 8
Battle Rhythms = Management

Level 7
One-On-One Meetings = Management

THE STRUCTURE
Build the how

Level 6
Individual Goal Setting = Leadership

Level 5
Department Goals = Management

Level 4
Accountability Charts = Leadership

THE FOUNDATION
Build the why

Level 3
Processes & Procedures = Management

Level 2
Standards & Expectations = Leadership

Level 1
Vision, Mission & Values = Leadership

Sixteen

Standards and Expectations — Where Values Meet Reality

Your staff don't need motivation. They need fucking standards. Set them. Live them. Enforce them. — Jack Delosa

We had a clear Vision, Mission, and Values. The flag was planted, and the language made sense to the team.

The next problem was execution.

Values give direction. They define what matters and what doesn't. But they don't organise behaviour. They don't define how a handover is done, what acceptable presentation looks like, or how people behave when no one is supervising the floor. That work sits with standards.

That's why standards sit where they do in the Operator System.

They're the layer that turns intent into something people can actually operate inside. Without them, values stay aspirational and execution becomes interpretive. People aren't acting against the values; they're just filling in the gaps themselves. Accountability becomes personal. Performance conversations turn subjective. Culture starts relying on personality instead of structure.

I didn't come into H&C unfamiliar with this idea. I'd been exposed to standards and expectations work through a leadership intensive with the Entourage in Melbourne in 2024, and I was already applying the same thinking inside Neptune. I'd seen how closely standards have to sit against values if either of them is going to matter.

What I hadn't done before was apply that system inside Defence.

Uniform, rank, inherited policy, and deeply embedded habits change how standards land. What works cleanly in business doesn't automatically translate onto the deckplate. If standards were going to work here, they had to be built in a way that fit the environment rather than fought against it.

Repeated Observation, One Conclusion

Two weeks after we rolled out our Vision, Mission, and Values, and in the lead-up to introducing the Standards and Expectations document, I made a point of walking through the galley just after 0630 most mornings. Breakfast service was already underway. Sailors moved through the line, plates in hand, fuelling up before the day took hold.

The space looked the way it usually did. Work moved. People knew their roles. Service continued without interruption.

What changed was how I watched it.

Across those mornings, patterns repeated. Presentation varied depending on who was on watch. Handovers carried different assumptions, which meant small issues were passed forward instead of being resolved. Each shift made sensible decisions in isolation, but the cumulative effect was inconsistency. The same work was being done, just not in the same way twice.

People weren't disengaged or careless. They were operating from habit and experience, because that was the guidance available to them. In the absence of something more specific, they filled the gaps themselves.

I'd given the team language for what mattered to us, but I hadn't yet given them a shared reference for how that language translated into daily behaviour. The intent was common. The execution wasn't.

Watching that play out across multiple shifts confirmed what I already understood. If the Operator System was going to work without me standing in the middle of it, values couldn't sit on their own.

The Standard I Set for Myself First

I didn't start by correcting the team.

I started by tightening my own execution.

I knew I couldn't define expectations for others until my part of the system was clean. If I expected people to arrive prepared, then the environment they walked into had to be prepared first. That responsibility sat with me.

My role wasn't to stand in the middle of the floor directing traffic. It was to make sure the floor could run without friction.

That meant earlier starts in the office, not the galley. Staffing issues were identified and resolved before they became last-minute scrambles. Training requirements were tracked properly instead of being remembered when something went wrong. Known problems were planned for instead of reacted to. Issues coming down from command were intercepted early, handled quietly, and resolved before they landed on the team.

The work was deliberate and mostly invisible.

What mattered was the result. The team walked into a space that functioned. They could focus on service instead of compensating for gaps upstream. The absence of chaos did more to lift performance than any instruction I could have given on the floor.

I also paid attention to how I handled my own misses. When something slipped at my level, I addressed it directly, fixed it, and moved on. No deflection. No excuses. The standard applied upward as well as down.

The team noticed long before anything was written. They watched how I showed up across different days, different workloads, and different moods. They watched whether my behaviour stayed consistent when things were inconvenient.

Once that pattern was established, expectations stopped feeling imposed. They felt credible.

Only then did it make sense to write them down.

Why I Tied Standards to Values

When I started writing standards, my first instinct was to list the obvious things. Arrival times. Uniform. Handovers. Clear rules that looked sensible written down.

The problem didn't appear on the page. It appeared the first time I had to correct someone.

When I framed an issue as a breach of policy, the response was predictable. The behaviour changed briefly, then returned to its previous pattern once attention moved elsewhere. The correction was registered, but it didn't last.

When I anchored the same correction to our values, the outcome was different.

Instead of pointing at a rule, I pointed at something the team had already agreed mattered. I explained why the standard existed and what it protected in the way we worked. The conversation stayed focused on the work, not on authority. Ownership followed because the reference point wasn't mine alone.

That mattered because it changed where accountability lived.

Once standards were written against values, they stopped sitting with me. People adjusted their own work earlier and with less intervention. Small issues were addressed before they accumulated. Correction happened inside the team rather than travelling upward.

One mechanism made this visible quickly.

When chefs began posting photos of their work in the group chat, presentation became clear without explanation. The standard was visible across watches. People compared their own output against what they saw and adjusted accordingly. Plates improved. Consistency followed. The detail started to matter again.

What stood out to me was where most of the standards originated. They surfaced through irritation. Meals that looked rushed. Handovers that set the next shift back. Humour that landed comfortably for one person and heavily for another.

Each moment pointed to the same issue. The values we'd named were stronger than the behaviour we were still accepting.

I focused on defining what good looked like in this environment, and why it mattered to the work. Standards were written as descriptions of execution, not warnings. The language reflected the job being done, not policy inherited from elsewhere.

That approach shaped everything that followed. Each recurring frustration became a prompt to define a standard. Each standard was anchored to a value the team already recognised. Over time, those references replaced interpretation.

That's what made the standards usable. People didn't wait to be corrected. They had something clear to work against.

The Grey Undershirt

About a week later, I noticed Ozzie wearing a grey undershirt that should have been replaced a long time ago. It was worn thin, stretched out, and visibly damaged. It looked like it had followed him through more deployments than it should have.

On its own, it was a small thing.

Ozzie was one of my senior leaders. People took cues from him without thinking about it. What he wore, what he tolerated, and what he overlooked quietly set a reference for everyone else on the floor. If I ignored it, I wasn't being flexible. I was redefining the standard without saying a word.

I didn't deal with it through a vague comment or an offhand reminder. That kind of feedback creates uncertainty. People make a token adjustment, then guess where the line actually sits.

Instead, I went back to the standards document.

Under Dress and Bearing, I added two lines:

Grey undershirts must be stain-free and free from visible wear. All uniform items will be maintained to a professional standard consistent with Pursue Greatness.

That was the entire change. No explanation attached. No expansion.

I pulled Ozzie aside, showed him the update, and explained it plainly. He was a visible reference point for the team. If his standard slipped, it told everyone else exactly how much the standard really mattered. This wasn't personal. It was about keeping the reference clear.

There was no argument. He understood immediately.

The next day, he turned up with a new undershirt. More importantly, he started holding his own people to the same expectation. The line didn't need reinforcing — it was written, visible, and shared.

That moment clarified something important for me.

The standard wasn't about clothing. It existed to remove ambiguity. Once expectations were written and tied back to values, accountability stayed factual. Conversations stayed short. Corrections stayed clean.

That's when standards stopped feeling discretionary and started functioning as intended.

How I Applied the Standards Inside a Large Organisation

When I started defining standards in H&C, I wasn't working inside a neat structure.

At HMAS *Coonawarra*, intent at the top was clear and deliberately broad. The Commanding Officer set three priorities for the year: look after our people, support the fleet, maintain base safety and security. Everyone understood them. No one argued with them.

They just sat a long way above the work.

What didn't exist was a usable link between that intent and the decisions being made inside my department during a normal shift. There was plenty of direction at the organisational level, but nothing a chef could apply during service, or a supervisor could apply during handover without interpreting what those priorities were meant to look like in practice.

That gap mattered. When intent sits above the work without translation, behaviour fills the space on its own. People act from habit and experience, not from shared expectation.

I didn't see that as a failure of Defence or command. I saw it as a consequence of scale.

A large organisation can set direction. It can't define behaviour for every function in detail. H&C operates under different conditions from engineering, logistics, or operations. Expecting one set of organisational expectations to guide behaviour consistently across all of that ignores how the work actually happens.

So I worked with the structure as it existed.

At the top sat organisational intent. That wasn't mine to rewrite. It set boundaries and purpose, but it didn't give instruction.

Inside the department, I had clarity. We knew why our work mattered, how we wanted to operate, and which behaviours we were prepared to defend. That was where standards could be shaped properly.

From there, expectations could be pushed closer to the work. Not as interpretations of policy, but as clear descriptions of how this department operated day to day.

I didn't try to build standards down from organisational language. I built them outward from the department and then into the teams doing the work.

I started with department standards.

These applied to everyone in H&C, regardless of role, rank, or shift. They described how people treated each other, how work was prepared and handed over, and which behaviours were acceptable. Their purpose was simple: remove ambiguity.

Once that baseline existed, it became obvious it couldn't stay generic.

Front of house and back of house worked under different conditions, carried different risks, and failed in different ways. Forcing a single set of standards across both didn't create consistency. It pushed judgement back onto individuals.

So the standards were refined further.

Department standards described how H&C operated as a whole.Front of house standards covered service flow, presentation, pace, and customer interaction.Back of house standards focused on preparation, hygiene, handover, and execution.

As expectations moved closer to the work, they became more specific. They stopped being something people agreed with and became something people could apply while the job was happening.

I didn't ask for permission to do this, and I didn't frame it as resistance. It was simply taking responsibility for the space I was accountable for and making sure expectations inside it were explicit.

Once that structure was in place, organisational intent stopped floating above the work. It showed up in service decisions, in handovers, and in how behaviour was corrected.

That was the only way standards could function inside an organisation of that size, and it matched the reality of the work my people were doing.

How the Standards Took Shape in Practice

Once I was clear where standards had to sit, the next question was what they actually needed to cover.

I didn't start with a blank page. I already knew the areas that mattered most to me in this department. The standards couldn't just describe tasks. They had to reinforce how people worked together, how the job was executed, how professionalism showed up day to day, and how capability was built beyond the next service.

Those became the four areas I focused on.

As the standards were written and applied, they were refined through observation. Not because the department was failing, but because clarity makes patterns visible quickly. The work showed me where definition mattered.

Some standards focused on how people interacted while the job was being done.

In a kitchen environment, small behaviours compound fast. Expectations around communication, conflict, and recognition existed to keep the team functional while the work was moving. They clarified how issues were raised, how disagreements were handled, and how effort was acknowledged, so none of it depended on individual judgement in the moment.

Another group focused on execution.

These covered handovers, preparation, hygiene, timing, and presentation. The intent was consistency across watches. Clear standards ensured the work was done the same way regardless of who was on shift, so people weren't making subjective calls about acceptable output mid-service.

Professional standards formed the next group.

These dealt with credibility. How people arrived, how prepared they were, and how they presented themselves to set a tone before any instruction was given. If we said we pursued greatness, those signals had to match the language. The standard was visible without explanation.

The final area was development.

If I was setting higher expectations, I was responsible for ensuring people had the capability to meet them. Training, mentoring, and progression were written into the standards. Senior people were expected to develop others. Knowledge was shared deliberately. Mistakes were examined and used.

I didn't name these categories at the start. I noticed them forming as the same types of issues kept surfacing and required the same kinds of correction.

The structure followed the work.

Once those areas were clear, expectations stopped needing explanation. Conversations stayed direct, accountability stayed factual, and the standards functioned as part of how the department operated rather than something separate from it.

Up to this point, everything in this chapter has been observational. What I saw on the floor. What I corrected. What changed when standards were clear and when they weren't. The question I kept getting from people after this work settled was straightforward: *how did you actually write the standards?*

How I Built the Standards and Expectations Document

I didn't sit down to design a standards document. I started by paying attention to what kept costing time, energy, or trust during a normal week.

The same things kept showing up. Handovers that set the next shift back before they'd even started. Presentation that depended on who was on watch rather than a shared expectation. People arriving technically on time but clearly not ready to work. Side conversations replacing direct ones when something needed to be said.

I didn't treat those moments as things to complain about. I treated them as signals that an expectation hadn't been written clearly enough.

For each one, I asked myself a single question: What behaviour do I actually want to see here, every time?

From there, I wrote the behaviour down as a standard.

Handovers are the simplest example. Instead of vague language about "effective communication between shifts", the standard described exactly what had to be in place before a handover was complete. What information had been passed on. What condition the workspace was left in. What responsibility transferred at that point.

The intent wasn't to catch people out. It was to remove the guesswork that kept forcing the next shift into recovery mode before they'd even started.

Presentation followed the same logic. Rather than leaving it to individual judgement, the standard framed presentation as an expression of respect for the people we were feeding and pride in the work itself. That wording mattered, because it tied the expectation directly back to values the team already recognised.

In practice, it meant standards around consistency, cleanliness, and care were written plainly enough that anyone could look at a plate and know whether it met the expectation or not.

Uniform and bearing followed the same logic.

After the undershirt incident, I stopped assuming there was a shared understanding and wrote the line explicitly. What was acceptable. What wasn't. Not in policy language, but in terms that reflected how we expected to show up as a department.

As the document grew, a pattern emerged. Every standard did three things. It described behaviour that could be seen. It explained why that behaviour mattered in our environment. And it was written tightly enough that it could be enforced without turning the conversation into an argument.

That last part was the filter. If I couldn't picture myself pointing to a line during a conversation and having it stand on its own, it didn't stay in the document.

Once the base standards were in place, I added a second layer deliberately.

These weren't about correcting poor behaviour. They were about lifting the bar where the team already had capacity.

Plating was one example. I wanted care and pride to be visible on the plate, not just food meeting the minimum requirement. I adjusted the standard so that it required at least one plated element at each meal service.

It didn't tell people how to plate. It told them that effort and attention were part of the job, not optional extras.

The same thinking applied to teamwork and development. Standards described how people stepped in when a teammate was under pressure, how knowledge was shared, and how training was treated as part of the work rather than something deferred until there was "time".

Before the document went anywhere near the team, I ran every standard through one final filter:

Was I prepared to enforce this consistently, including when it was uncomfortable?

If the answer was no, it came out.

Only after that did I involve the team. Not through a formal rollout or briefing, but through direct conversations. I walked people through the standards as living expectations, explained why they existed, and made it clear they applied to everyone, including me.

That's how the document took shape.

Not as a rulebook dropped from above, but as a written version of how the department already operated when it was at its best. Tightened enough that no one had to guess where the line sat.

If you want to see what that looks like in practice, an example of the Standards and Expectations document is available on the Frontline Horizon website under Resources. It's there as a reference point, not something to copy, but a way to see how clarity was written directly into the work.

Enforcement: Making Standards Real

Writing the standards wasn't difficult. Living them was.

Once the document existed, there was no room for selective application. Every standard only mattered if it applied when it was inconvenient. The first time I chose to let one slide, I would have been teaching everyone exactly how optional they were.

What I learned quickly was that enforcement didn't require intensity. It required timing and consistency. Issues addressed early stayed small. Left alone, they blended into the background and quietly became normal.

How those conversations were handled mattered. When feedback sounded like policy, people shut down. When it referenced a standard we had already agreed to, the conversation stayed anchored to the work. The standard carried the weight, not my position.

That only worked because the standards applied to everyone. Rank didn't change the expectation. Performance didn't change the expectation. I didn't give myself an exemption either.

Over time, enforcement stopped feeling like a separate act. It became part of the working rhythm. Expectations were corrected in the moment, without drama, because the line was already clear.

The conversations that made that possible followed a pattern.

I didn't invent it. I learned it from Shane Byrne, my mentor at the Entourage, and then pressure-tested it repeatedly in environments where mistakes had consequences and relationships mattered.

Eventually, that pattern became the structure I relied on whenever a standard was missed or exceeded.

We called it S.O.F.A.

Once I started using it consistently, accountability stopped feeling confrontational and started feeling fair.

That's where we go next.

S.O.F.A. MODEL
THE OPERATOR'S CONVERSATION FRAMEWORK

PHASE 1
S – SITUATION
WHAT HAPPENED

PHASE 2
O – OUTCOME
THE RIPPLE EFFECT

PHASE 3
F – FEELINGS
EMOTIONAL TRUTH

PHASE 4
A – AGREEMENT
FORWARD COMMITMENT

Corrective Conversations Only
BELIEF STATEMENT
"I KNOW YOU'RE BETTER THAN THIS"
Styled as an additional tactical element for corrective situations only.

Seventeen

Bonus Section: The Critical Conversation Framework — S.O.F.A. Model

Catch your staff doing the right thing 51 percent of the time, "You catch more flies with honey than vinegar". — Ken Blanchard and Spencer Johnson (one minute Manager)

I first came across the idea years ago on the Leading Seaman Promotion Course in 2010. *The One Minute Manager* was required reading at the time, and one concept stayed with me long after the course finished.

The idea itself was simple. If you want behaviour to repeat, you have to notice it while it's happening and name it clearly. Not after the fact. Not buried in a report. In the moment.

What stuck wasn't the book. It was how sharply the idea cut across what I was seeing in uniform.

On paper, we say we value people doing things right. In practice, most attention goes to what's wrong, what's missing, or what's about to fail. Some of that focus is necessary. Complacency does real damage. But when correction becomes the only time your people hear your voice, something else starts to erode quietly in the background.

Trust thins out. Engagement drops. Pride fades.

I've watched it happen too many times to ignore. People stop associating leadership with growth and start associating it with pain. They brace when you walk toward them. Over time, even good people stop lifting, because there's no upside to being noticed.

Somewhere along the way, I stopped overthinking it and settled on a rule I could actually live by when the tempo was high.

Catch your people doing the right thing more often than you catch them doing the wrong thing.

Not because mistakes don't matter. They do. But when most interactions are corrective, the system breaks long before the work does.

That extra one percent matters because it forces intent. It makes you lift your eyes from the problems and notice when someone steps up, helps a mate, lifts a standard, or quietly does the right thing without being asked. When those moments are named properly, you don't just make people feel good. You reinforce behaviour you want repeated.

That links directly to the reason values came first in meetings and training. Values set the reference. They explain what matters. But values only shape behaviour when they're reinforced in real moments, not just discussed in theory.

Outside of those group settings, I needed a way to handle the conversations that actually shape culture on the floor. The one-on-one moments. The corrections. The recognition. The pauses where something needs to be said.

What I realised was that the same structure I used to correct behaviour cleanly could also be used to reinforce excellence without it turning into hollow praise. Using one framework for both kept feedback balanced, grounded, and real. People didn't have to guess whether a conversation was going to hurt or help. They knew it would be deliberate.

S.O.F.A. gave me that balance.

This isn't part of the Operator System, and I'm not presenting it as a universal solution. It's simply a framework that worked for me and one I still rely on. It gives structure to moments that are easy to mishandle. It keeps me steady in the conversation and stops me slipping into rank, ego, or frustration when the pressure is on. Most importantly, it ensures the other person leaves knowing exactly what happened, why it mattered, and what good looks like.

What matters is that the structure stays the same every time. I don't change it depending on whether the conversation is positive or corrective. The framework doesn't shift. The intent does.

That consistency is what makes it work. People don't have to guess what kind of conversation they're walking into. They know it will be clear, direct, and worth listening to, whether they've nailed the standard or missed it.

Over time, I started paying attention to how and when I gave feedback. The pattern wasn't intentional, but it was consistent. The correction had structure and weight. Praise was looser, often improvised, and easier to brush past. Once I noticed it, I couldn't ignore it.

That imbalance was mine to own.

I didn't want my people guessing what it meant when I called them over. I didn't want praise to feel casual and correction to feel heavy by default. I wanted every conversation to land with the same clarity and intent, whether someone had met the standard or missed it.

That's why the structure never changes.

S.O.F.A. stays the same every time.

What changes is the direction in which it's applied.

Situation

This part of the conversation is about describing what actually happened. It is grounded in what I saw, not in interpretation or assumption.

In a positive conversation, that means naming the behaviour I want to encourage — where someone stepped up, took ownership, or lifted the standard without being prompted. The point is to make the behaviour visible and unmistakable so it can be repeated.

In a corrective conversation, it means identifying the behaviour that created the issue. What was done, or not done, that led directly to the situation we are now dealing with. Nothing more than that.

The rule is the same either way. I avoid judgement, labels, and commentary on character. The moment a conversation starts with something like, "you're disrespectful," it stops being about behaviour and turns into an argument about identity. People defend labels far more aggressively than they fix actions.

Starting with "this is what happened" keeps the conversation anchored in reality. Reality is harder to deflect, harder to debate, and easier to correct.

Outcome

This part of the conversation is about the ripple effect. It is where behaviour is connected to impact beyond the moment it occurred.

In a positive conversation, that means spelling out what their action enabled. How it helped the team, supported the mission, reduced pressure, or made the next shift's job easier. It also includes naming the less obvious impact, such as how their behaviour lifted the standard for people watching and quietly set a reference for what "good" looks like.

In a corrective conversation, the same clarity applies in the opposite direction. I'm explicit about the cost. Who was affected, what was delayed, what was compromised, and where the pressure landed as a result. Not in an exaggerated way, and not as a threat, but as a factual account of consequences that already occurred.

What I've learned is that people adjust behaviour faster when they understand outcomes that land on real people. They don't change because authority was challenged or rules were enforced. They change when they can see how their actions helped or hindered the people around them.

Feelings

This is the part of the conversation where honesty matters most, and where I had to be disciplined with myself.

I'm not talking about venting, theatrics, or unloading emotion because I've had a bad day. I'm talking about calm, deliberate emotional truth, stated plainly and kept under control.

In a positive conversation, that might mean telling someone you're proud of how they handled a situation, genuinely appreciative of the effort they put in, or impressed by the way they stepped up without being prompted. Those words only matter when they're tied to specific behaviour and delivered without exaggeration. Empty praise is easy to ignore. Honest recognition isn't.

In a corrective conversation, the tone changes but the principle doesn't. You state the emotional truth plainly. Disappointment. Concern. A clear sense that the standard wasn't met. Said once, owned, and left there.

When it's delivered calmly and without heat, that honesty lands harder than raised voices or formal warnings. Not because it threatens consequences, but because it makes the impact clear while keeping the conversation intact.

I've learned that when someone you respect tells you they're disappointed, calmly and directly, it lands harder because it makes the impact personal without turning the conversation hostile.

Agreement

This is the point where the conversation turns forward and the standard is made explicit. It's about commitment.

In a positive conversation, this is where you lock in what right looks like. You reinforce that the behaviour you've just acknowledged is the expectation for the future. You're not praising a one-off moment. You're confirming that this is how the team operates and that you expect to see it repeated because it matters.

In a corrective conversation, the commitment is just as clear, but it points in a different direction. You agree on what will change next time and describe what acceptable execution looks like under pressure. The focus isn't on what went wrong. It's on what will happen instead.

I always want that agreement spoken out loud. Not nodded through and not assumed. Said clearly, so both of us leave the conversation with the same understanding of what's expected.

The Belief Statement (Corrective Conversations Only)

This is the part I'm most deliberate about, and the part I don't use casually.

I don't include it in positive feedback. Belief is already implied there. If I'm acknowledging someone for doing the right thing, they already know I back them. There's no need to underline it.

I reserve this for corrective conversations, and I finish with it every time.

"I know, and you know, that you're better than this."

Said properly, calmly, without sarcasm or threat, that line carries weight. Not because it cuts someone down, but because of what it does at the same time. It holds them accountable for what happened, makes it clear they're not being written off, and reinforces that they still belong at the standard we've set.

If you mean it, people feel that immediately. It lands as belief, not pressure. It challenges them to lift, not because they're afraid of consequences, but because they don't want to let down someone who still backs them.

That distinction mattered to me. I saw the difference play out over time. Correction that creates short-term compliance looks clean in the moment but rarely sticks. Correction grounded in belief produces growth, because it keeps people engaged in the standard instead of pushing them away from it.

S.O.F.A. was never meant to be a script or a lecture. It became a structure I could rely on when the moment mattered. Once I'd run it enough times in my head, I didn't need notes or preparation. I could step into a conversation calmly, even when I was tired, frustrated, or caught off guard, and still handle it deliberately instead of reacting on instinct.

Here's how it looked in real time.

A Corrective Conversation: The Handover Failure

This one mattered because it didn't just look sloppy. It created immediate disruption.

The morning shift walked into unsecured equipment, dirty prep areas, and cold storage left unlocked. That isn't a minor oversight. That's safety, quality, time, and trust taking a hit all at once.

I spoke with the Leading Seaman in charge and kept the conversation focused.

I started by naming what had happened. When the morning shift arrived, equipment was unsecured, prep areas hadn't been cleaned, and cold storage was left unlocked.

Then I made the impact clear. Prep was delayed, the morning team started behind the line, and service was under pressure before it even began. More than that, it sent the message that handover was optional instead of a standard the team could rely on.

I was direct about how it landed with me. I told him I was frustrated and disappointed, not because of a single miss, but because this was a standard we had already discussed and agreed to.

Then I moved the conversation forward. I asked for agreement that from now on, handover would be personally verified before the shift ended, with a report and photos sent through to the POML-C every time. No exceptions.

I finished the way I always do in a corrective conversation.

"I know you're better than this."

That line matters. It keeps the correction from turning into shame. Shame makes people hide, and hidden problems always come back bigger. Belief keeps the conversation constructive and keeps the person engaged with the standard instead of checking out of it.

That's what the framework looks like in practice.

A Positive Conversation: Doing It Right

I used the same structure when someone got it right.

LS Lemont took initiative with a new joiner. There was no instruction from me and no checklist driving it. She saw a gap and stepped in without being asked.

I called it out the same way.

I started by naming what I'd seen. That morning, I watched her pull the new sailor aside and walk him through the handover properly instead of letting him work it out on the fly.

Then I explained the impact. Because she did that, the next shift started clean. There were no delays, no confusion, and no recovery work needed. She saved the team time and stress, and she helped someone feel like they belonged from day one.

I was clear about how it landed with me. I told her I was proud of the way she handled it and genuinely impressed by the initiative she showed. That behaviour was leadership, whether it came with a rank badge or not.

Then I locked it in. I asked for agreement that this was how we brought new joiners up to speed and that I expected her to keep setting that example, because it mattered to how the department ran.

Same structure. Different direction.

She walked away knowing exactly what she did, why it mattered, and that it wasn't just a nice moment. It was part of the standard.

That's what S.O.F.A. looks like when it's lived, not explained.

What Makes This Work in the Real World

S.O.F.A. isn't a shortcut, and it doesn't replace courage. You still have to step into the conversation when it would be easier to walk past it. What it does remove is the mental load.

When I'm tired, frustrated, or short on time, the structure keeps me from drifting into vague criticism, rank-driven discipline, or emotional reactions I later wish I'd handled differently. It anchors the conversation to what happened, why it mattered, how it landed, and what comes next. That consistency matters, especially when pressure is high.

There was another realisation that took me longer than it should have.

In a corrective conversation, the reason I'm there is never mood, stress, or irritation. It's always because of something that happened, or didn't happen, against a standard we've already set. Keeping that clear changed how I showed up. The conversation stayed where it belonged, on the behaviour and its impact, not on me. When I held that line, defensiveness dropped away because I wasn't making it personal. I was responding to reality.

Over time, I noticed how people reacted to that consistency. Even when they were being corrected, they knew what was coming. They knew the conversation would be direct, controlled, and grounded in the work. They also knew they weren't being written off. That predictability built trust in moments that usually damage it.

That's why I came to treat S.O.F.A. as a tool, not a theory. I was never interested in polished explanations or leadership language. What mattered was having something I could rely on when the moment tested me, when the conversation was awkward, when I was tired, or when avoiding it would have been easier.

S.O.F.A. didn't make those conversations comfortable, and it wasn't meant to. It gave me a way to step into them without hiding behind rank, ego, or emotion. It kept my focus on what happened, why it mattered, and what needed to change next.

OPERATOR SYSTEM

THE MASTERY
Lead like a pro

THE EXECUTION
Build the rhythm

Level 9
Daily Systems = Self Management

Level 8
Battle Rhythms = Management

Level 7
One-On-One Meetings = Management

THE STRUCTURE
Build the how

Level 6
Individual Goal Setting = Leadership

Level 5
Department Goals = Management

Level 4
Accountability Charts = Leadership

THE FOUNDATION
Build the why

Level 3
Processes & Procedures = Management

Level 2
Standards & Expectations = Leadership

Level 1
Vision, Mission & Values = Leadership

Eighteen

Processes and Procedures — Systemise Everything

A bad system will beat a good person every time.   — W. Edwards Deming

There's a point in every team turnaround where the energy is real; the values are clear, the standards are agreed, and people are leaning in. You can feel it. Conversations change. Plates leave the pass looking like someone actually cared. The banter comes back, but it isn't cruel. The place has a pulse again.

Then reality taps you on the shoulder.

Because standards don't execute themselves.

You can be crystal clear on what matters and still watch things unravel on a random Monday morning. You're short-staffed. The roster's ugly. Someone's called in sick. A new joiner is standing there trying to work out what's expected, and the pace doesn't slow down to help them. That's the moment where intent either holds or collapses.

That's where processes and procedures come in.

I'm a big believer in them, not because I enjoy structure for its own sake, but because when they're done properly, you buy time back as a leader immediately. Dan Martell talks about this in *Buy Back Your Time*, which is one of the most practical books I've read on building leverage. The idea is simple. If the system can carry the work, the leader doesn't have to.

In the Operator System, this is the how layer. Processes and procedures are what turn expectations into something repeatable under pressure. They're the bridge between what you say matters and what happens when conditions aren't ideal.

Done properly, they don't slow down teams. They create trusted speed. When people have a clear blueprint, they stop hesitating, stop double-checking, and stop waiting for permission. They move faster because they're certain. That certainty is what allows ownership to take hold without the leader becoming the bottleneck.

That's the reason processes and procedures sit in the Foundation of the Operator System. If this was the only layer you added on top of values and standards, you'd still see a disproportionate return for the effort involved. Clarity increases. Rework drops. Questions slow down. The same expectations get met more consistently, even when you're not standing there watching it happen.

This is the final piece of the Foundation. It's the last layer of Build the Why before we move into structure, rhythm, and mastery.

Processes and procedures aren't about inspiration. They're about continuity. They keep the standard alive when conditions aren't perfect. They stop the department relying on a handful of high performers to hold everything together. Most importantly, they give new people a fair shot at doing the job properly from day one, without embarrassment, confusion, or being set up to fail.

Navy Does This Better Than Most People Want to Admit

I've spent a lot of my life inside military systems, and I'll say something that usually surprises people, including those of us who've worn the uniform long enough to criticise it properly. The Navy does procedures reasonably well.

Not always elegantly. The paperwork can be clunky. The language is often overcooked. Some documents clearly haven't been written by someone who's done the job recently. But the instinct underneath it is sound.

In the Navy, most critical work has a process. Most recurring tasks have a routine. Most handovers have a protocol. That isn't about controlling people or stripping away initiative. It exists because the environment punishes inconsistency quickly and without mercy.

On a ship, a lazy handover isn't a minor inconvenience. It becomes a safety issue. A missed check isn't an "oops". It can turn into a crisis at sea. You don't get the luxury of learning lessons slowly when the consequences arrive immediately.

Procedures exist because you don't get to choose the conditions you operate in. The work still has to be done whether you're fully crewed or short, whether the watch is experienced or green, whether the person stepping into the role joined last week or last year. The system has to hold even as the people change.

That discipline stayed with me long after I stepped outside Defence. It's also where I noticed the biggest gap the moment I started working with civilian organisations.

The Civilian Gap: Hope Is Not a System

That gap showed up for me when I started working with the Entourage and being exposed to other people's businesses. I'd sit in rooms listening to owners describe recurring problems, bottlenecks, and frustrations, and then I'd start asking basic questions about Standard Operating Procedures, quick reference guides, checklists, or delegated authority. More often than not, I'd get blank stares. It felt like I'd switched languages halfway through the conversation.

In civilian workplaces, I saw a lot of teams operating on what I eventually started calling *hope and heroes*. Hope that people remember how things are meant to be done. Hope that knowledge gets passed on properly. Hope that Sarah is around, because Sarah knows the thing.

And there is always a Sarah. Or a John. Or a Leila.

She becomes the unofficial training package and the walking memory bank. She's the person who quietly fixes problems and keeps the wheels on without ever making a fuss about it. Everyone relies on her, even if no one admits it out loud. The system appears to work, but only because she's holding it together.

Until she's sick. Or she leaves. Or she burns out and starts resenting the place, because she's carrying knowledge that should have been written down years ago.

When that happens, everything starts to wobble. Deadlines get missed, mistakes appear out of nowhere, and quality drops. Everyone acts shocked, like the failure came out of the blue.

It didn't.

The system was never real. It was just people holding things together with experience, goodwill, and gaffer tape. The moment the people changed, the cracks became obvious.

I've written more than a hundred SOPs for Neptune Holdings Group. That's not a flex, it's context. Even after writing so many, I still had moments where something went sideways and my first reaction was the same one I'd had in uniform. Why is this happening again?

The answer was usually simple. Either the process was poorly designed, or the process was sound and people weren't following it, or both at the same time.

That distinction matters, because those are different problems with different fixes. And that's the real diagnostic this chapter is built around. Not theory or best practice, but working out whether you've built a system, or whether you're still running on hope and heroes.

Process Issue or People Issue

The language around "process problems versus people problems" isn't something I invented, and it didn't come from one neat quote either. The thinking traces back to W. Edwards Deming, the American statistician whose work shaped modern quality, safety, and operational leadership. Deming argued relentlessly that most failures don't come from bad people, they come from bad systems. His view was that the majority of errors sit with the process and the environment people are working in, not with individual effort or intent.

That idea has been absorbed into military, safety, and operational cultures over decades. The wording has changed, but the principle hasn't. Before you blame a person, you need to understand the system they were operating inside.

In my experience, when something goes wrong operationally, you're almost always dealing with one of three situations.

The first is that the process itself is broken. It might be unclear, unrealistic, outdated, or never properly defined. People are guessing, filling gaps with habit, or doing what makes sense to them under pressure.

The second is that the process is sound, but people aren't following it. The steps exist, the expectations are clear, but enforcement is inconsistent, training is weak, or ownership isn't obvious.

The third is the worst version. The process is broken, and the team has stopped respecting it altogether. At that point, no one follows it, everyone does their own version, and the system quietly collapses into chaos disguised as flexibility.

If you don't work out which one you're actually dealing with, you'll burn weeks solving the wrong problem. You'll blame people when the process is rubbish, or you'll keep rewriting procedures when the real issue is accountability and culture.

I've made both mistakes.

I've assumed someone was lazy or careless, only to realise we'd never actually shown them the right way to do the job. We'd expected them to pick it up through observation, trial and error, and whatever they could infer on the fly. That isn't development. It's gambling.

I've also gone the other way and rewritten the same SOP multiple times because I thought the document was the issue, when the real problem was that no one believed the standard mattered enough for it to be enforced.

That's why, before I touch a pen, there's one question I force myself to answer honestly.

Is this a process problem, or a people problem?

If you can't answer that clearly, you're about to create busy work instead of fixing what's actually broken.

The 'More Than Twice' Rule

Like the idea of process versus people problems, the 'more than twice' rule doesn't belong to a single person. It comes out of lean systems thinking, long before it had a catchy name. The principle shows up in the Toyota Production System, in continuous improvement cultures, and later in productivity and scaling work across military, safety, and business environments.

The logic is simple. When a task repeats, variation creeps in unless the work is deliberately designed. Repetition is a signal. If something keeps happening, it's telling you it wants structure.

I first heard it articulated clearly by Dan Martell in one of his talks. He called it the 'more than twice' rule, and it stuck with me because it gave language to something I was already doing instinctively. If I had to explain something more than twice, that was my cue to stop relying on memory and build a system.

Dan explained it using an image that landed immediately. If you had to paint a wall with a hundred birds on it and you painted every bird by hand, you'd be slow, careful, and exhausted by the end. But if you took the time once to build a stencil with ten birds, you could finish the wall in minutes. Same outcome. Far less effort. Far more consistency.

That's the rule in plain terms.

When a task shows up more than twice, it deserves a system. Something that can be followed consistently, regardless of who's doing the work or how familiar they are with it. The goal isn't documentation for its own sake. The goal is repeatability.

That system doesn't need to be heavy. It might be a checklist, a short written procedure, a visual prompt in the workspace, or a quick reference guide. It could be a short video recorded once and turned into a simple SOP. The format doesn't matter. What matters is that the next person doesn't have to reinvent the job from scratch.

Once repetitive work is systemised properly, a few things happen almost immediately. The standard holds, even when the person changes. Delegation becomes real, because you're no longer handing over your memory or experience, you're handing over a process. Errors reduce, because simple checks catch what tired brains miss. And you get space back, because you're not answering the same questions on repeat.

That last part is more important than most leaders realise. People burn out not because they're working too hard, but because they become the default answer for everything. Every interruption feels small but stacked together they drain focus and judgement.

A process is the stencil. It carries the repetition so your brain doesn't have to. That's what frees you to do the work only a leader can do, instead of acting as a walking instruction manual.

What Needs Procedures and What Doesn't

One of the easiest traps with procedures is trying to document everything. The volume grows fast, people tune out, and the system starts to feel like admin instead of support.

The better question is where a procedure actually earns its place. For me, it's simple: write procedures for work that repeats, work that carries real consequences when it's missed, and work that gets messy when people are new, tired, or moving fast.

That usually points to the same areas. Onboarding, because the first week sets the standard more than any speech ever will. Daily routines, because that's where consistency is either built or quietly lost. Safety, because risk management is culture in action. Quality checks, because standards only stay real when they're verified. Communication, because clarity stops small issues turning into big ones. And anything that kicks in under pressure, where you want a response you can trust without needing a leader in the middle.

That's where procedures do their best work. They take recurring friction and turn it into something reliable.

What This Looked Like in a Navy Galley

H&C isn't a ship's control room, but the operating conditions are closer than most people expect. Meal service is repetitive, high tempo, and unforgiving. Breakfast doesn't get delayed because a handover was loose, and two hundred sailors don't wait patiently because the system is having an off day. Food still has to move. Hygiene still has to hold. Temperatures still matter. Headcount still has to be managed. Service still happens on time.

Because of that reality, procedures were never optional. They weren't just about survival; they were about pride. If we were serious about excellence, the work had to be repeatable. It couldn't depend on who happened to be on shift, who was tired, or who had more experience on a particular watch.

It became clear quickly that procedures weren't there to control chefs. They were there to protect them. A well-built procedure doesn't restrict creativity. It protects the non-negotiables so creativity has somewhere stable to sit. Certain parts of the job simply can't fail. Temperatures, hygiene, storage, allergen controls, cleaning routines, and handover integrity don't change based on mood or talent. They stay fixed because the consequences of getting them wrong are immediate and serious.

Once those foundations were locked in, everything else had room to move. Plating could evolve. Flavours could improve. Presentation could lift. Service energy could change from shift to shift. That's where pride lives, and that's where good chefs want their attention focused.

We built procedures around what could not fail and deliberately left space everywhere else. The system carried the load, and the people focused on the craft. That's what freed them. Not the absence of structure, but the right structure in the right places.

The Mistakes I Kept Seeing When Procedures Failed

I once heard a combat story that stuck with me.

A unit had a simple drill. Whenever they exited their armoured vehicle, they exited on the left. It was clean, rehearsed, and automatic. Under fire one day, they did exactly that.

The problem was the rounds were coming from the left.

The drill worked perfectly. The judgement didn't.

That story captured the real tension with procedures. SOPs are essential, but when they remove thinking at the point of execution, they stop being protective. Discipline isn't blind compliance. Discipline is knowing the standard and still being able to assess and adapt when conditions change.

That distinction shaped how I looked at procedures everywhere else.

Over time, I noticed the same pattern whenever procedures failed to land. It was rarely about effort or attitude. Most of the time, the document itself couldn't be used when it mattered.

Defence is a good example of this tension. We recognise the need for procedures, but we often write them like legal briefs. Long, dense, and built on the assumption that everyone reads and absorbs information the same way.

When a procedure takes too long to read, people default to habit. Under pressure, they do what feels familiar.

Damage Control procedures showed this clearly for me. These are the systems that protect the ship when things go wrong. I was an experienced sea-going sailor, yet I rarely read the SOPs in full. Not because they weren't important, but because they didn't reflect how the work actually unfolded in real conditions.

What we trained wasn't the document. We trained behaviour. The procedure existed, but the real system lived in repetition and muscle memory.

That gap matters. When a document isn't usable, standards drift. Not through defiance, but because the procedure doesn't help in the moment it's needed.

The procedures that get used are the ones people can follow when they're tired, busy, or filling in on an unfamiliar shift. They're clear, practical, and written like one competent person explaining the job to another.

That led me to a simple test.

If a new person can pick up the procedure and execute the task properly without rescue or workaround, the system is doing its job. If they can't, the problem isn't the person.

It's the procedure.

Good SOPs standardise what must be consistent, protect what can't fail, and still leave room for judgement. Anything else is just paperwork pretending to be a system.

How I Build Procedures Without Them Turning Into a Bureaucratic Nightmare

Whenever I take over a department, one of the first things I do is review the existing operational procedures. Not because I assume the previous leader got it wrong, but because it's the fastest way I know to understand how the place actually runs. Procedures reveal what the team believes matters. They show where risk is being managed, where problems have occurred before, and where friction tends to appear.

I don't stop at reading the documents.

I watch the work. What people say they do and what actually happens on the floor are rarely identical. Paper can look clean while reality tells a different story. Watching the job unfold shows where people hesitate, where they improvise, and where the system starts to strain under normal operating pressure.

Over time, I learned that the cleanest way to build or repair a process is to start at the edges, not the middle. I always begin with triggers and handovers. What starts the process. What ends it. Who hands it over. How that handover happens. And what 'good' looks like at that exact point. Most operational breakdowns don't live inside the task itself. They live in the space between people.

Once I understand that flow, I don't sit in an office and write procedures in isolation. I bring in the key operators, the people who do the work every day, and ask them to walk me through the reality of the job. Not the ideal version. The real one. The shortcuts, the workarounds, and the things that only make sense once you've done it enough times to feel where it bends. That context is the difference between a process that reads well and one that works.

From there, I draft the procedure myself. I keep that part centralised because consistency of language and structure matters. Once it's written, it goes straight back to the team for pressure testing. We walk through it step by step and ask simple questions. Does this work in practice? Does this step make sense at pace? What fails if we follow this exactly as written? The document gets adjusted until it reflects how the work is done, not how we wish it was done.

The final test is the one that matters most. I hand the procedure to someone who has never done the task before and ask them to follow it section by section without help. I watch where they pause, where they hesitate, and where the document assumes knowledge they don't yet have. If someone new can follow the procedure and achieve the same outcome as someone who runs the task instinctively, the system is sound.

That's the standard I hold procedures to. Not perfection, but reliability. A process that works regardless of who's on shift, how tired they are, or how long they've been in the role. That's when procedures stop feeling like bureaucracy and start functioning as infrastructure.

How Procedures Become Part of Culture

Writing procedures is straightforward. Getting them embedded into daily work is where most teams struggle. A procedure doesn't become real because it exists in a folder or gets signed off in a meeting. It becomes real when it's trained, used, referenced, and updated as part of how the work actually runs.

In H&C, whenever we changed how service was run, we didn't rely on verbal updates and hope they stuck. We updated the checklist. We briefed it properly. We practised it on the floor. And we made sure the next shift didn't have to rely on someone else's memory to get it right. That's how consistency is protected before small deviations become normal.

We were just as deliberate about where procedures lived. They weren't buried in shared drives or hidden behind multiple logins. They were kept close to where the work happened. Printed. Visible. Easy to access. Easy to reference while the job was moving.

That mattered, because access determines use. If a procedure takes effort to find, it won't be used when time is tight. That isn't a discipline issue. It's human behaviour. We built systems that worked with how people operate, not how they work against it.

When procedures are used properly for training, kept close to the work, and reinforced through use, they stop feeling like documentation and start shaping behaviour. That's the point where they move from being written instructions to becoming part of the culture.

The 1-3-1 Method

Where Problems Become Processes

This is one tool I use to turn recurring problems into procedures. Dan Martell outlines it in *Buy Back Your Time*. The labels are his, but the behaviour is what good operators do when they want a team to think, not just report issues upward.

I've seen the opposite pattern my entire career in uniform and in business. A problem gets dropped on the leader with no thinking attached. "Chief, the oven's broken." "Boss, the client's unhappy." "Sir, we've got an issue." If that's where it ends, the leader becomes the default brain of the operation, and the team stays dependent.

1-3-1 changes the input. One clearly defined problem. Three viable options. One recommendation. The point isn't to be clever; it's to make sure the person bringing the issue has already done enough thinking to move the situation forward.

Where it becomes useful in this chapter is what happens next. If the same type of problem keeps showing up, and a 1-3-1 conversation produces a solution that works repeatedly, that's your signal to systemise it. That's when the fix earns a procedure.

And I don't write it for them. The person closest to the problem usually has the best context to build the first draft. My role is to pressure-test it, tighten it, and make sure the standard is clear.

Inside the Operator System, 1-3-1 is a filter. It keeps problems from becoming leadership traffic, and it helps you spot which recurring issues deserve to be turned into processes the team can rely on.

A Real Neptune Example: Why SOPs Still Fail

Even with procedures in place, I still saw breakdowns in business. On paper, the SOP existed. The document was clear. And the work still wasn't happening the way it should.

The issue was never the words on the page. It was ownership.

Inside Neptune, whenever a procedure wasn't being followed, it usually came back to one of three things. No one was clearly responsible for owning it, so accountability blurred and everyone assumed someone else had it covered. The staff hadn't been trained in the process properly, so when pressure increased, people defaulted to habit. Or the procedure was sound, but there was no follow-up, so ignoring it carried no weight.

All three point to the same gap. The SOP existed, but no one truly owned it.

That doesn't get fixed by rewriting the document. It gets fixed by making ownership explicit and placing it with the right person.

As Neptune grew, this became impossible to ignore. Just because I'm the CEO doesn't mean I should be training a Security Officer at Palmerston Shopping Centre on a procedure I've never personally executed. That responsibility belongs with their direct supervisor. The person who does the work, understands the pressure points, and can coach it properly in real time.

My responsibility is different. I make sure that supervisor knows the procedure exists, understands they own it, and is accountable for training it consistently until it becomes normal. Ownership follows responsibility, not title.

There are exceptions. When the SOP is onboarding a new Executive Leadership Team member, that sits with me. It's work I'm directly involved in and accountable for, so I own the process end to end. The rule stays the same. Ownership sits with the person closest to the work and the one responsible for its outcome.

Once that line is clear, procedures stop being documents and start shaping behaviour. People know who owns what, where to go for clarity, and what happens when expectations aren't met.

That's also why the next layer of the Operator System matters. Processes only work when ownership is visible, and visible ownership is what turns expectations into something enforceable. That's where accountability charts come in — where we're heading next.

The Quiet Power of Procedures

Procedures don't make you a hero. They create stability. They make standards repeatable, delegation practical, onboarding cleaner, and day-to-day work easier to execute under pressure. When the tempo is high, they replace guesswork with clarity.

When I look back at when H&C genuinely turned around, procedures were never the headline. Values were visible. Standards were clear. Culture felt different. But underneath all of that, procedures were what allowed the change to last.

Even when I wasn't on the floor, the work kept its shape. Shifts followed a familiar rhythm. Handovers had a baseline. Quality checks happened often enough to keep standards alive. Not perfectly, but consistently enough that things didn't unravel the moment attention shifted.

That was the real marker. The system didn't rely on strong personalities or constant leadership presence. The work continued in a way that reflected the standard we'd set.

Once you reach that point, leadership changes. You stop carrying the work and start directing it. The system holds, the team moves, and your energy goes where it actually matters.

Where to Start Without Overcomplicating It

I'm not going to walk you through a full drill here, because that belongs somewhere you can use it. What I do want to do is give you enough context to understand what you're building and why it matters before you go and do the work.

If you've read this chapter and you're thinking, *alright, I get the idea, but where do I actually start*, the answer is simple. Start with one recurring task that matters. Something that happens often, carries risk when it's done poorly, or creates friction when it's inconsistent. That's the kind of work where procedures earn their keep.

The mistake people make at this point is either trying to document everything at once or choosing something abstract because it feels important. Don't do that. Pick something real. Something you see, something you've had to correct, something that keeps coming back to you as noise.

Once you've identified it, the work isn't about clever wording or perfect formatting. It's about understanding how the task actually runs today. Where it starts, where it finishes, and where it gets handed over. That understanding becomes your blueprint.

I've laid out a practical drill under **Resources on the Frontline Horizon website** that walks you through this process step by step.

It's designed to help you identify what needs to be systemised, how to document it without turning it into bureaucracy, and how to test it so it works under real conditions.

Where This Takes Us Next

Processes and procedures are how standards survive contact with a messy day. Values set direction. Standards define the line. Procedures make the line repeatable when you are not standing there.

Once procedures exist, the next question is unavoidable: ownership. Who owns them, who uses them for training, who updates them, and who steps in when they're missed.

That is why the next layer matters. Accountability charts make ownership visible, so the system can run without everything routing back to the leader.

OPERATOR SYSTEM

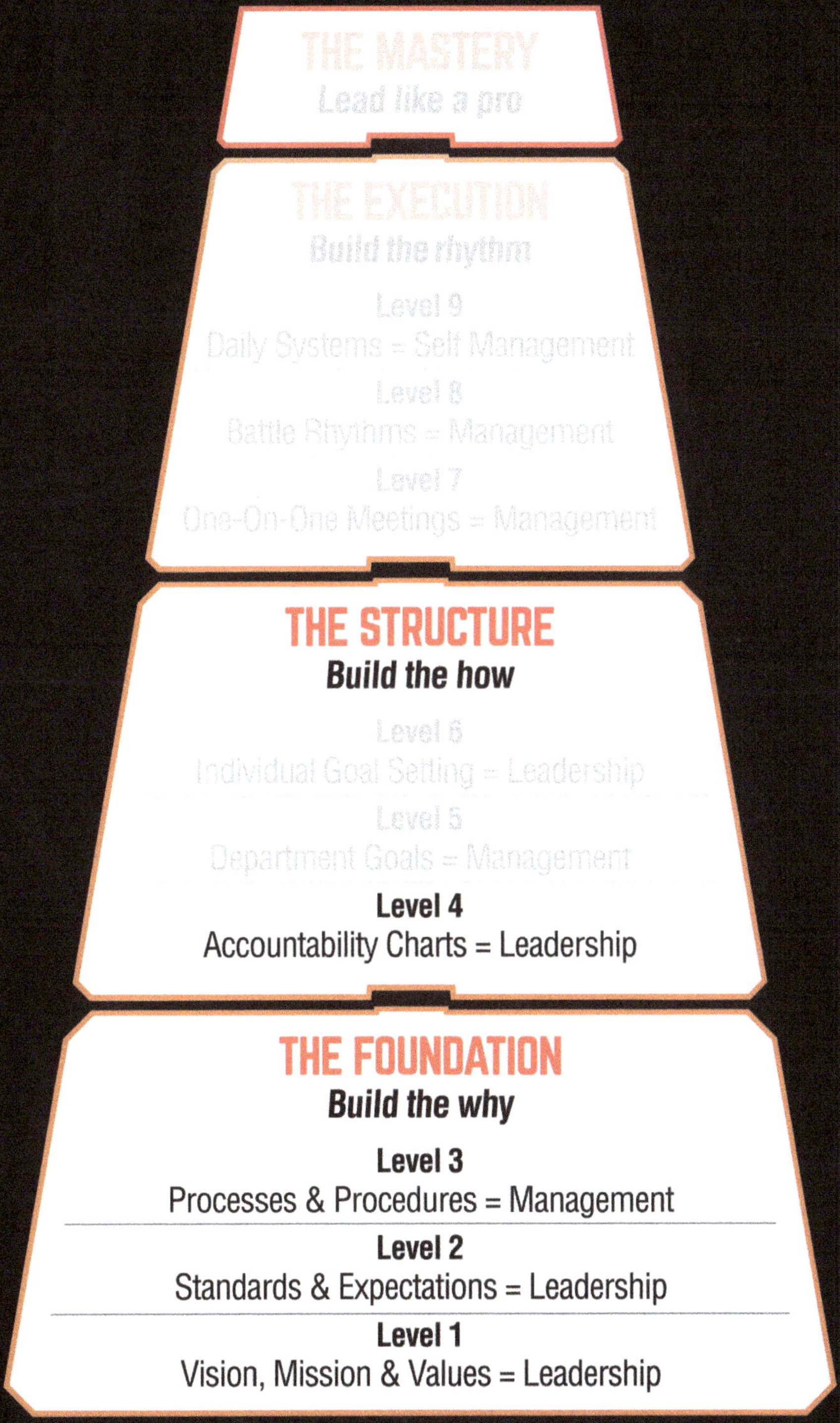

Nineteen

Accountability Charts — Who Does What

Accountability is not about blame. It is about clarity.   — Unknown

You walk into work and nothing is broken. The place is running. People are moving. Tasks are getting ticked off. From the outside, it looks fine.

Then you start following things up.

You chase something you assumed was handled. You ask who owns it and get different answers. Someone thought it was done. Someone else thought it sat elsewhere. No one is avoiding the work. No one is dodging responsibility. No one can point to a single name and say, that sits here.

You feel it first in the drag. Decisions take longer than they should. Work gets repeated or missed. Background tasks drift unless you are watching them. Every time you step away, something slips just far enough to need another correction.

Over time, people slow themselves down. They hesitate before acting because ownership is unclear. Acting carries risk, so they check first. They wait. The system keeps moving, but it leans on you more than it should.

You end up carrying work without doing it. You hold the place together through presence and follow-up. The team is capable. The structure is unfinished.

That was the frustration accountability charts solved for me. Not by adding oversight, but by making ownership visible enough for the work to sit where it belonged.

That pattern was already familiar when I stepped into H&C. Because of that, it was something I moved on early. Once the foundation was in place, the next move was accountability charts, settled deliberately and fast.

Before I started using accountability charts properly in 2021 and later in Neptune, this is how departments typically operated. Progress appeared whenever someone had capacity. Training advanced when the right person had time. Safety stayed sharp while it had attention. Events came together with late effort. Watches ran, but adjustments depended on who noticed first. Nothing collapsed outright, but everything relied on presence rather than structure.

That created a predictable rhythm. Things held when I was close to it. When I took leave, cracks appeared. Decisions bottlenecked back to me. Questions returned that never should have needed my input. Ownership was not anchored tightly enough for people to act with confidence.

That is where accountability charts came in.

I was not introducing something new to myself. I was applying a tool I already trusted to an environment where the work was already defined. The aim was simple. Stop responsibility drifting and let the system carry itself.

Once ownership was mapped against the functions that actually mattered, decisions landed where they belonged. Background work held without prompting. I stopped being the reference point for work that no longer needed to come back to me.

That is what accountability charts changed in practice. They did not alter the work. They clarified who carried it.

This is also where accountability charts closed a gap that formal position descriptions never addressed.

This is where Navy earns a mention. Position descriptions are treated like paperwork. Generic language. Broad responsibilities. Plenty of words. Very little day-to-day clarity. They tell you your rank and trade, but they rarely tell you what you actually own when you walk in each morning.

Accountability charts work from the opposite direction. They ignore rank and focus on function. They answer a practical question that position descriptions avoid. Who owns this when it needs attention, when it slips, or when it needs to be improved.

Once functions were mapped instead of titles, ownership became visible straight away. Training stopped floating. Safety had a name next to it. Galley control no longer depended on who was loudest or most senior on the shift. Events stopped turning into scrambles. Decisions landed where they belonged instead of drifting back to me.

That shift did not require another brief or explanation. It required structure.

This chapter covers how I used accountability charts to make ownership visible, why they cut through confusion faster than policy ever did, and how they became the hinge point between systems that look good on paper and systems that actually hold under pressure.

Where This Clicked for Me

I first came across accountability charts in *Rocket Fuel* by Gino Wickman and Mark C. Winters. What the book gave me wasn't a new idea. It gave me a practical way to make something I already understood visible to a team without relying on rank, volume, or personality to force it through.

The value was visibility.

Seeing accountability laid out against functions instead of titles made it usable. It turned something I had been carrying implicitly into something I could point to, reference, and operate from without hovering over the system to keep it moving.

That clarity is why accountability charts sit where they do in the Operator System.

By the time you reach this layer, the work is already defined well enough that people know what good looks like. What comes next is structure that allows responsibility to sit with the people closest to the work without constant input from you.

In any build, the person in charge does not do every trade. Each trade owns its piece of the job completely. That internal ownership does not remove accountability from the person in charge. It sharpens it.

This is something Jocko Willink captures well with extreme ownership, and it matches how I've operated in practice. Inside the team, accountability can and should sit with those doing the work. Outside the team, accountability never moves. If something fails, it sits with the leader. That responsibility is carried and protected, especially when mistakes are made.

This layer sits where it does in the Operator System because everything that follows depends on ownership being explicit. Once responsibility is anchored to names and functions, direction becomes usable, goals become ownable, and department focus operates without relying on you. Accountability charts are the structural element that allows the system to carry load without you propping it up.

Why Organisation Charts Don't Answer the Right Question

In the military, job titles are clear. Leading Seaman (Maritime Logistics Supply). Petty Officer (Maritime Logistics Chef). Able Seaman (Bosun's Mate). Those titles tell you trade and rank. They tell you where someone sits in the hierarchy.

They do not tell you who owns the work that keeps a unit functioning day to day.

Every unit has functions that matter and don't sit neatly inside a position description. Someone has to own the Ship's Safety and Environment Team. Someone has to make sure welfare events actually happen, not just get discussed. Someone has to restock and check first aid kits. Someone has to own communal cleaning routines, training records, and general stores that don't clearly belong to one department. When those things aren't owned, they stop getting done properly, and you usually find out at the worst possible moment because sailing without toilet paper is real shitty situation.

The civilian world has the same problem with different labels. Customer support floats between sales and operations because there isn't a dedicated support role. CRM data degrades because everyone's busy and it's treated as shared responsibility. Onboarding gets patched together because HR is one person covering too much ground. Supplier relationships stay vague until a delivery fails and everyone starts scrambling.

An org chart doesn't solve any of that. It was never designed to. It shows authority and reporting lines. It does not make ownership visible. When something gets missed, the impact lands on the people closest to the work.

There is a simple rule underneath all of this. When everyone is responsible, no one is accountable. The moment ownership is shared by default, responsibility becomes optional. Accountability charts exist to remove that ambiguity by forcing ownership to be explicit, singular, and visible.

That isn't a people problem. It's a structure problem.

What Accountability Charts Change on the Ground

When accountability charts are in place and used consistently, teams operate with clarity.

Authority sits where the work sits, so decisions are made without delay. Responsibility is defined tightly enough that people know exactly what they own and where their function begins and ends. Information flows to the right places because ownership is visible. Outcomes are carried by names, not activity.

Accountability charts turn day-to-day operation into something predictable. Meetings stay focused because decisions already have owners. Work moves forward because action does not require permission. Leaders stop being the collection point for unresolved work and start operating at the level they are meant to.

A good accountability chart creates momentum. Ownership is obvious. Decisions move. Work progresses without follow-up.

The Three Principles They Are Built On

I stopped treating accountability charts as something you design and started treating them as something you uncover. The work already exists. The chart simply makes ownership visible once you're honest about what needs to be carried and who is positioned to carry it.

Everything I've built with accountability charts rests on three principles. When accountability breaks down, it's because one of these has been compromised.

1. **Functions come before names.** I list the work that needs ownership before I think about who fills the role. That includes the work that usually lives in the gaps between job titles. If a function matters enough to cause friction when it slips, it matters enough to be named. This step exposes how much of the operation is being carried by assumption rather than structure.

2. **Right people in right seats.** Ownership follows capability, interest, and capacity, not seniority or time served. The best owner of a function is the person who can carry the outcome consistently. When the fit is right, issues surface earlier and action happens without chasing.

3. **Single-point accountability.** Each function has one owner. One name carries the outcome, even when execution involves others. This keeps responsibility anchored and prevents the chart from degrading into something that looks organised but doesn't operate under pressure.

When any of these principles are diluted, the chart stops doing its job and ownership no longer operates cleanly under pressure.

Two Rules That Saved Years of Friction

Every accountability chart I've ever built rests on two rules. When these are applied cleanly, ownership stays intact. When they're compromised, responsibility drifts.

The first rule is that accountability follows the role, not the task.

If a function sits with someone and they choose to delegate part of the work, that's fine. Delegation is part of leadership. What does not move is ownership of the outcome. If the function is not delivered to standard, the person named on the chart owns the result, regardless of who they passed tasks to.

If your name sits next to the function, the outcome sits with you.

That applies in uniform, and it applies in business. If an Operations Manager owns onboarding and delegates pieces of it to a junior coordinator, and onboarding fails, the failure belongs to the Operations Manager. Sharing execution does not dilute accountability.

The second rule is to assign functions based on capability, not seniority.

I saw this clearly on HMAS *Leeuwin* in 2018. We had a Seaman onboard with a Master's degree in Environmental Science. In practice and on paper, they were the most qualified person on the ship to contribute meaningfully to the Ship's Safety and Environment Team. They had the background, the interest, and the credibility to lift the standard well beyond basic compliance.

They weren't even given the opportunity because all anyone could see was the fucking rank.

Instead, the responsibility went to more senior people who had neither the background nor the interest to do more than the minimum. The result wasn't failure. It was mediocrity. A box-ticking program that met the requirement and went no further, while the person who could have improved it sat on the sideline.

That made the trade-off obvious. Prioritising seniority protects hierarchy at the expense of performance. Matching functions to people based on capability produces better outcomes and develops judgement where it's actually needed.

A Real Navy Accountability Chart (How This Looked in Practice)

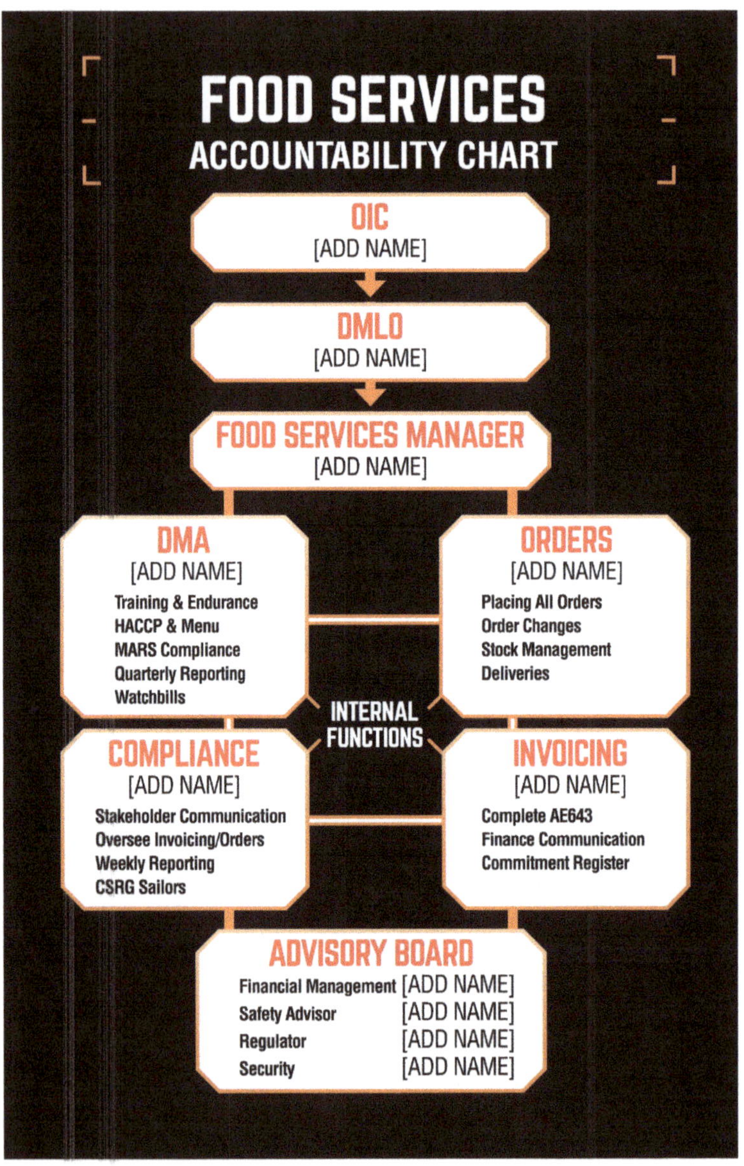

When I was the Food Services Manager, this is how accountability worked day to day. The department wasn't organised around rank or trade. It was organised around functions that needed to operate consistently.

My role was clear. I owned overall performance, stakeholder engagement, resourcing, standards, and final decisions when issues crossed functions. Below that, Food Services was broken into internal functions that reflected how the work actually moved.

Department management and administration had an owner. Orders and supply chain had an owner. Compliance had an owner. Invoicing had an owner.

That clarity changed the way the department operated. If a delivery was wrong, it went straight to the owner. If the commitment register wasn't current, it didn't sit in limbo. If compliance checks weren't being actioned, the conversation didn't turn into a post-incident argument. It went to the function owner and stayed practical.

Those owners didn't do every task themselves. Orders still involved the team. Compliance still required coordination. Admin still depended on information flowing in. What changed was that responsibility landed cleanly. When something slipped, the question was not who was meant to do it. The question was what broke and what needed to change so it didn't repeat.

That is what accountability charts provide when they're built around functions instead of titles. Responsibility is visible before something goes wrong, not assigned after the fact.

How I Correct Misalignment When I See It

When an accountability chart isn't working, I don't question effort or attitude. I assume the structure is off and I go looking for where ownership and authority are misaligned.

If ownership overlaps, I force a decision. We define the function in plain terms, including what good looks like, where the work starts, and where it finishes. Then I assign one owner. Others can support the work, but responsibility sits with a single name. Once that's clear, hesitation disappears.

If authority sits without accountability, I reset the boundary. Anyone making decisions that affect outcomes carries the result. If they're not prepared to own it, they don't keep the authority. That usually means shifting accountability to match the authority, or pulling authority back to where accountability already sits.

If accountability exists without authority, I treat it as a system fault. I look at what's missing and fix it. Access, information, permission, tools, or backing. If the person named as owner cannot influence the outcome, either the authority needs to be added or the ownership needs to move. Leaving someone accountable without control is how you burn capable people.

When these adjustments are made, I make them visible. Ownership shifts are named. Authority changes are stated plainly. Not for effect, but so the chart reflects how decisions are actually made.

If the same issue resurfaces twice, the chart is the first place I go. Repeated problems usually mean ownership is unclear, overloaded, or assigned to the wrong function.

Making It Work Day to Day

An accountability chart only matters if it is used in daily operation.

When a decision was needed, we didn't default to rank, volume, or availability. We went to ownership. If your name sat next to the function, the call landed with you.

When tension appeared between people, we didn't start with intent or personality. We went to ownership. This function sits here. That function sits there. This is what good looks like. It gave us a neutral place to resolve issues without making them personal.

As conditions changed, the chart changed with them. When workload shifted or someone became overloaded, ownership was adjusted deliberately so the structure reflected how the work was actually being carried.

When something didn't happen, the response stayed practical. We went to the owner to understand what broke and what needed to change so the standard was met next time.

Ownership meant exactly that. If you owned the function, you owned the result. Support existed and context was considered, but responsibility did not drift back to me by default.

How I Build Accountability Charts

I treat accountability charts as a discovery process. The goal is to expose where ownership exists, where it's assumed, and where it's missing, then make that visible enough to operate from.

I start by listing the functions, not the people. Before I think about names, ranks, or personalities, I write down everything that needs ownership for the team to operate properly. That includes the obvious work and the background tasks that usually only get attention when they fail.

In a Navy environment, that list usually includes safety and environment coordination, welfare activities, first aid management, training records, general stores, communal cleaning routines, watchbill coordination, equipment logs, and the integrity of handovers. None of those sit cleanly inside a single trade or rank, but all of them matter to how the unit actually runs.

In business, the functions change, but the principle stays the same. Customer support, invoicing, vendor management, CRM hygiene, onboarding coordination, reporting, compliance, scheduling, asset control. If it affects customers, cash flow, safety, or reputation, it goes on the list.

Once the functions are clear, I assign one owner to each. Ownership and execution aren't the same thing. One person owning the outcome doesn't mean they do all the work. It means they are accountable for making sure it is delivered to standard.

I don't list co-owners or backups on the chart. People can support the work and be trained as relief, but the chart needs to stay clear. Two names against a function creates overlap, and overlap kills urgency.

After that, I make it visible and I use it. When something needs a call, we go to the owner. When something is stuck, we go to the owner. When something keeps slipping, we go back to the function and the ownership and work out what's misaligned.

That's when accountability charts stop being a concept and start functioning as infrastructure.

HOW TO BUILD ACCOUNTABILITY CHARTS GRAPHIC 10

Why This Layer Changed Everything for Me

Once accountability was explicit, the Operator System started to work the way it was designed to.

Ownership sat with the work. Decisions were made by the people closest to it, with full understanding of context and consequence. Responsibility came with authority, so action moved without permission needing to be sought.

The biggest change was what no longer required my involvement. The system no longer depended on my presence to stay upright. Decisions that didn't need me stopped reaching me. Loose ends stayed where they belonged.

That didn't make the environment perfect. It made it dependable. Work progressed without constant intervention, and outcomes no longer relied on me being everywhere at once.

Once accountability is in place, the structure behaves as intended. Values are lived without reminder. Standards are applied consistently. Processes are followed without relying on memory or heroics. Ownership stays anchored.

From there, the focus can shift to direction.

With ownership established, effort can be pointed deliberately toward department priorities. Focus becomes clear. Movement aligns with outcomes that matter, not activity that only looks busy.

This is where the next chapter begins.

OPERATOR SYSTEM

THE MASTERY
Lead like a pro

THE EXECUTION
Build the rhythm

Level 9
Daily Systems = Self Management

Level 8
Battle Rhythms = Management

Level 7
One-On-One Meetings = Management

THE STRUCTURE
Build the how

Level 6
Individual Goal Setting = Leadership

Level 5
Department Goals = Management

Level 4
Accountability Charts = Leadership

THE FOUNDATION
Build the why

Level 3
Processes & Procedures = Management

Level 2
Standards & Expectations = Leadership

Level 1
Vision, Mission & Values = Leadership

Twenty

Department Goals — Sprinting toward Your North Star

There is nothing so useless as doing efficiently that which should not be done at all. — Peter F. Drucker

I'd seen versions of this problem for most of my career before I ever had language for it. Teams that functioned. Departments with solid people, standards mostly understood, workable processes, and ownership documented. On the surface, nothing looked wrong. Work was getting done, tasks were completed, meetings happened, and emails were answered.

The issue wasn't dysfunction. It was stagnation.

I've watched Naval Officers and senior sailors walk into work, open their computer, and spend the entire day inside their inbox. Questions answered. Requests handled. Issues closed out. By the end of the day, the inbox is empty, and it feels like a productive day because no one ever showed them another way to measure progress.

At close of business, the department looks much the same as it did at open. Capability is unchanged. Standards haven't fallen, but they haven't moved either. The work has been done, and in most Defence and government environments, that's enough. "Work getting done" becomes the finish line.

In business, maintaining the same position isn't neutral. Customer expectations change. Markets evolve. Competitors improve. What was acceptable six months ago stops being enough, even if nothing external forces the issue. If progress isn't deliberate, the organisation settles into a version of "good enough" that requires the least effort to maintain, and momentum disappears.

In my experience, this is a direction problem.

When an organisation, department, or team doesn't have an explicit destination, the work defaults to whatever arrives first and demands attention. Responsiveness fills the day. The manager stays busy and reactive and starts to feel like they're running on a treadmill. A lot of movement, no forward travel.

This is where the next layer of the Operator System comes in.

Up to this point, the work had been about stability. The department could function without constant correction. It just wasn't moving.

What came next was direction. If the department was going to move, not just function, it needed something to move toward.

I was about six weeks into H&C. The foundations were in place. Behaviour was consistent. Standards were holding. Processes were running without constant correction. Accountability was anchored. You could stop there and, by Defence standards, call it a job well fucking done.

But I hadn't come in to just hang around.

From the start, I had a clear BHAG in mind. A Big Hairy Audacious Goal is a long-term, deliberately bold objective that's clear enough to judge and big enough to force real change. In my case, it wasn't a slogan or a performance metric. It was proving the Operator System framework in a real Navy environment, end to end, under normal constraints.

To do that, stability wasn't enough. The system needed a direction.

That's where department goals came in. Ninety-day sprints that translated a longer-term ambition into something the team could act on. Concrete outcomes. Clear ownership. Effort aimed at things that genuinely moved the needle.

At that point, I could have settled. I could have ridden out the rest of the posting, kept things ticking over, and walked away discharged and satisfied. Plenty of people do exactly that. I'm not built for that kind of ending.

This was the layer that changed what the team was capable of.

Department goals took a team that could already execute reliably and gave that execution intent. Effort was aimed, energy focused, and work started moving in a deliberate direction. In my honest, biased opinion, that's what turned H&C into the highest-performing Navy department in 2025.

A department can operate cleanly every day and still remain static. Direction is what creates movement.

That's what department goals provide. They set where the department is heading next and give you a clear way to see, in real time, whether the work is moving it forward or simply filling the day.

What OKRs Get Right and Why They Didn't Fit Navy

I want to give credit where it's due. The thinking underneath this layer isn't something I invented.

In business, most people recognise it as OKRs — Objectives and Key Results. The language is modern, but the logic isn't. Drucker was writing about management by objectives decades before dashboards existed. Grove operationalised it at Intel. Doerr scaled it at Google. What people now call OKRs is a refined version of a much older idea.

I learned OKRs through the Entourage world, specifically through my mate Andy Mac, and they were useful. They forced discipline around outcomes. If you can't clearly state what you're trying to achieve and how you'll know when you've achieved it, you're not directing effort. You're managing motion.

That principle stands.

Where OKRs start to break down isn't the thinking. It's the environment they assume.

Proper OKRs rely on clean, frequent measurement. In business, that's realistic. CRMs, dashboards, live data, automated reporting. Progress can be reviewed weekly without anyone rebuilding spreadsheets or chasing numbers manually.

Most Navy departments don't operate like that. We're stuck somewhere between 1990 and 2005, using Excel like a priest uses a bible.

What you usually have instead are whiteboards, shared drives, manual reports — don't get me fucking started on the amount of reports, and routines layered on routines, with people already stretched thin. When you try to run pure OKRs in that environment, they turn into admin. Updates become the work. Reporting becomes the burden. The system meant to create focus starts competing with the job itself.

I didn't throw the idea away. I stripped it back.

In Defence, I stopped calling them OKRs and started calling them department goals or ninety-day sprints. Same intent. Simpler structure. More execution.

This is also where KPIs sit, because people regularly confuse these tools and try to force one framework to do every job.

KPIs tell you how the system is performing right now. They're health indicators. They show whether standards are being maintained, whether output is stable, and whether something is drifting. They're most commonly used at the individual level, but they can be applied at team or department level when the data supports it.

OKRs work well when the infrastructure supports them. They suit environments that can track progress rigorously and want a tight link between ambition and measurement.

Department goals aren't a midpoint between frameworks. They're a translation. They carry the same outcome-driven discipline in a form that works without heavy data systems. They're easier to communicate, easier to track manually, and easier to enforce without turning the system into a reporting exercise.

If I'd had clean data pipelines and modern tooling in Navy, I could have implemented OKRs more formally. But leadership isn't about loyalty to frameworks. It's about imposing structure that survives the environment it's dropped into.

For me, Department goals were the cleanest way to achieve the outcome.

Different operators will choose different tools at this layer of the Operator System. The label is irrelevant. What matters is whether the system creates clarity, focus, and forward movement without becoming another thing the team has to service.

The Difference Between a Destination and a Busy Week

A department can be well trained, well run, and working hard, and still not be moving anywhere. The work happens. Systems operate. People stay busy. But without a clear point the effort is aimed toward, there's no shared way to judge whether any of that work is actually improving the department.

When that reference point is missing, activity increases but progress becomes harder to identify.

At this level, management needs two things operating at the same time.

The first is a clear outcome. A practical picture of where you want the department, business, or organisation to be over the next twelve to twenty-four months. In Defence, that window is usually shaped by posting cycles. In executive roles, it may extend further. Either way, it needs to be concrete.

Not an aspiration. An outcome you can describe.

It should be obvious how the department operates differently, which standards are fixed rather than negotiable, and what observable change would tell you the department is in a better place than it is now.

That longer-term outcome matters, but it is not something most teams need to carry day to day. Pushing a twelve- or twenty-four-month horizon onto an operational or junior-heavy team tends to create distance rather than alignment. It becomes abstract and disconnected from the work that fills the week.

That is why I never tried to push the full picture down the chain all at once.

Instead, I held the longer-term direction myself and translated it into ninety-day sprints.

A sprint is the second element you need. It is a defined, time-bound block of work that answers a much simpler question: what are we working on right now that moves the department closer to where it needs to be.

This mirrors the same logic as Vision, Mission, and Values. You hold the broader intent through the vision. The sprint functions like the mission. It takes a larger aim and turns it into something executable in the current environment.

The longer-term direction sets where the department is heading. The sprint defines the next stretch of ground to cover. Both are necessary, but they operate at different levels. One provides orientation. The other drives execution.

In Navy, direction defaults unless you deliberately set it. Postings rotate, people change, priorities shift, and the tempo never pauses while you decide what you're trying to achieve. If you don't choose a twelve- to twenty-four-month direction on purpose, the department starts taking its cues from whatever lands in the inbox first or rolls downhill from above. The work still gets done and everyone stays busy, but movement stops being intentional. Over time, the department drifts wherever it's pulled.

A clear example of this for me was HMAS *Leeuwin* in 2018.

We were in rough seas. Properly rough. For Navy readers, think sea state five and building. For everyone else, heavy swell, spray everywhere, and weather that made being outside a bad decision.

Then the pipe came over the broadcast:

"All spare hands muster flight deck for fresh water wash down."

For civilians, a fresh water wash down is something you do before returning to port. You hose salt off the upper decks to slow corrosion. It's a sensible job. A necessary one. Just not in the middle of a storm where the ocean is actively throwing salt at you faster than you can remove it.

So there we were on the flight deck, in rain and sea spray, hosing down a surface that was being re-salted in real time by the weather. If you've ever seen that image of someone trying to mop the ocean, that was us.

Busy. Wet. Working hard. Achieving absolutely fuck all.

The issue wasn't effort, and it wasn't even intent. The task was clear. What was missing was alignment between the work being done and the outcome it was meant to produce in that moment. The activity made sense in isolation, but it didn't move anything forward.

That's the failure department goals are designed to prevent.

They force you to test whether the work you're asking people to do actually advances the outcome you care about right now. They create a reference point that lets you ask whether something is worth time and energy today, rather than defaulting to habit or tradition.

Without that filter, teams stay busy and committed, but effort gets spent on work that looks right while delivering nothing meaningful. With clear goals in place, activity has to justify itself against progress.

That's the difference between motion and movement.

The Levers Are Different, but the Logic Is the Same

Over time, I realised that moving any organisation forward always comes down to levers. The environment changes, but the logic doesn't. If you want progress, you need to know which inputs actually create movement and then pull them deliberately.

In business, I think about this in very simple terms. If an activity is not tied to revenue generation or profit optimisation, it is secondary. You can dress things up with strategy language, but in practice only a small number of levers actually move the organisation forward.

In my experience, they are these three:

- Lead generation or marketing, which creates opportunities

- Sales conversion, which turns opportunities into revenue

- Delivery at scale, which ensures the work can be done consistently without the system breaking.

You can argue nuance around the edges, but if those three aren't working, the business doesn't move. Everything else exists to support them.

In Navy, the levers are different.

Most departments are not responsible for attraction or recruitment, and they are not closing sales. That part of the equation sits elsewhere. But that doesn't mean departments lack leverage. It just means the levers that matter live in different places.

For the vast majority of Navy departments I've worked in, progress comes down to three things:

- People

- Training

- Delivery.

That's it.

Those are the levers that determine whether a department merely functions or actually improves. Department goals exist to focus effort on these levers deliberately, instead of pulling them at random or only after something breaks.

The sections that follow break each one down in practice.

People as a Performance Lever

People come first whether we like it or not. How you onboard people, develop them while they're with you, and send them out the door when they post or move on shapes performance more than any process ever will.

People who feel ignored, stuck, or invisible don't sustain their best effort for long. Not because they're lazy, but because nothing in the system is giving that effort somewhere to go.

Under Brigg at FLSE-D, direction was clear even when the tempo wasn't. Plenty of things landed in his inbox that could have become "today's emergency", but he filtered noise, protected the team, and kept us aimed at what actually mattered. Watching that up close taught me something simple: you can be busy all day and still not move the unit forward.

After he posted out, I saw the opposite. When direction isn't explicit, the day defaults to whatever arrives first and demands attention. You end up living inside other people's priorities and calling it productive because the inbox is empty and the meetings happened. That's how standards soften and capable people switch off without anyone noticing until it's already happened.

Training as a Growth Lever

Training is the second lever, and it's where I've seen the biggest gap between what organisations say they value and what they actually build.

In practice, most training I experienced early in my career was about coverage. Learn the process. Get signed off. Be competent enough to operate safely. That approach keeps the system running, and sometimes that is exactly what's required. But it rarely changes how people think, decide, or show up under pressure.

That's not what I mean when I talk about training as a lever.

When I ran departments, I treated training as development. Work that changed how people made decisions, how they handled pressure, how they took responsibility, and how they saw themselves in the system. Leadership, judgement, safety beyond rules, resilience, accountability, and basic mindset. Not instead of trade skills, but alongside them.

I deliberately exposed my people to training that wasn't strictly in trade. Not to turn sailors into executives, but because I wanted them to have skills I wish someone had given me at those ranks. In my experience, better humans become better operators. Broader perspective improves judgement. People who can see a future for themselves engage more deeply in the present.

One of the most important sessions I ever ran came from a mindset exercise I learned through Entourage, taught by Johnnie Cass. It wasn't technical and it wasn't procedural. It focused on the internal stories people carry and how those stories quietly shape behaviour.

I'll use LS Ozzie here, because that moment mattered.

At the time, Ozzie had come off the back of a brutal twelve months. Anyone who's been through a rough posting knows how that kind of stretch can distort how you see yourself. During the session, we were talking about internal narratives, and Ozzie shared the story he'd been telling himself. His words were blunt. He believed he was a cunt. Unapproachable. Hard to deal with. Not someone people wanted to be around.

The hardest part was that he believed it.

I asked him if I could run a quick exercise.

I turned to the room and asked, "Put your hand up if you believe that story about Ozzie is true."

Not a single hand went up.

Then I said, "Put your hand up if you believe the opposite is true."

Every hand went straight up. Twenty-five sailors. No hesitation.

I asked Ozzie to stand up and look around the room.

He didn't speak at first. When he turned back toward me, it was written all over his face. He'd been carrying a story that wasn't true, and because he believed it, it had been shaping how he showed up every day.

That single reframe changed everything.

From that point on, Ozzie operated differently. More confident. More engaged. More willing to step forward instead of holding back. By the end of 2025, he was one of the highest-performing sailors in the department, not because we taught him another process, but because we removed something that had been quietly limiting him.

That's what I mean when I say training is a lever.

In my experience, people don't disengage because the work is demanding. They disengage when they feel stuck. Training that only helps someone do today's job a little better keeps them where they are. Training that helps them grow beyond their current role gives them momentum.

That momentum shows up everywhere. Standards improve. Safety behaviour matures. Initiative increases. Retention improves because staying feels like progress, not stagnation.

This is also where department goals and training link cleanly. When you know where you want the department to be in twelve to twenty-four months, you can train toward that future deliberately instead of reacting to gaps after they cause damage.

I never saw training as a cost to be justified. I saw it as infrastructure that supported every other outcome we cared about. When training lifts how people think and operate, delivery becomes easier, standards hold with less effort, and goals stop feeling aspirational and start becoming achievable.

Delivering on Capability

The third lever is delivery, and this is where structure starts paying dividends.

Delivery is about how work actually moves through the system. Where friction appears, where duplication hides, and where capable people burn energy on tasks that don't create value but somehow never get questioned.

I'll give you a blunt example.

As a senior sailor, I had reporting requirements like everyone else. That part was normal. What wasn't normal was writing essentially the same report four different ways for four different parts of the organisation, all of which eventually rolled up to the same senior officer further up the chain.

That isn't accountability or governance. It's the system creating work to justify itself.

All it would have taken was one person upstream asking a simple question: what decision does this actually inform, and what changes if it exists versus if it doesn't?

Instead, the burden sat downstream. Time was burned rewriting the same information. Attention was pulled away from work that mattered. People stayed late not because the job was demanding, but because the system was inefficient.

Optimising delivery isn't about cutting corners or lowering standards. It's about designing the work so effort goes where it creates value. Remove duplication and unnecessary reporting, and capacity reappears almost immediately.

When delivery isn't optimised, you don't just waste time. You drain capability. Good operators end up feeding the machine instead of improving outcomes, and no amount of motivation or intent fixes that.

Why This Matters for Department Goals

This is where department goals stop being a planning exercise and start directing real work.

Once you understand which levers you can actually pull in your environment, goals give you a way to apply them deliberately. They let you decide what matters now, where effort should go, and how progress will be judged, instead of letting the day set those priorities for you. An operationally heavy quarter in a Fleet Support Unit might mean most of the focus sits on delivery. A quieter period in Fleet Logistics might allow you to lean harder into training. The levers stay the same. The emphasis shifts.

Without clear goals, the levers still exist. People are onboarded. Training happens. Delivery continues. But it drifts. Effort spreads. Training becomes reactive. Delivery settles into whatever keeps the system running. Everyone stays busy, but no one can clearly articulate what the department is actually building toward.

Department goals bring those levers into alignment.

They give managers context beyond today's task list and give supervisors the why behind the work. Training becomes preparation for a future state instead of a compliance activity. Delivery tightens because work is measured against outcomes, not motion. The same levers are still being pulled, but now they reinforce each other instead of competing for attention.

That's when effort starts producing momentum instead of just filling time.

What a Clear Goal Looks Like Inside a Team or Department

In business, long-range goals often get expressed as revenue, margin, customer growth, retention, or delivery metrics. Sometimes they drift into vanity measures when people lose the plot views, clicks, followers, numbers that look good on a slide but don't actually feed anyone.

I still remember a line from Jack Delosa that stuck with me: *what's the point of $100M in revenue if you spent $99M getting it?* That isn't anti-growth. It's a reminder that numbers without context are a trap.

In large organisations, including Navy, the same rule applies. A meaningful goal doesn't need to be complex. It needs to be real, time-bound, and observable.

At the department or team level, that usually means a twelve-month goal you can clearly judge. Not a slogan. Not a vibe. Something you could point to at the end of the year and say, *yes, this happened* or *no, it didn't.*

For example:

- "Within twelve months, this department will be recognised as one of the safest in the organisation, measured by a sustained reduction in incidents, clean audit results, and safety behaviours that hold without supervision."

- "Within the next year, we will be competitive for the Silver Platter based on independent assessment, not internal optimism."

- "Over the next twelve months, this will be the department other units benchmark for training quality and delivery consistency."

- "By the end of the year, a new joiner can walk into this department, get up to speed quickly, and succeed without being smashed or carried."

Those are not motivational statements. They are direction statements. They describe a future state clearly enough that you can work backwards from them.

Where most teams fail is stopping there.

A twelve-month goal on its own doesn't move anything. It only becomes useful when you translate it into what has to happen next. That's the role of ninety-day sprints.

The annual goal defines where you're heading. The ninety-day sprint defines what you are building right now to close the distance. Without that translation, goals stay abstract. With it, they turn into something the team can execute against immediately.

That's why this layer doesn't end with a goal statement. It ends with a sprint.

Why Ninety Days Worked for Me

I landed on ninety-day sprints because they fit how people actually work and how organisations actually behave.

Twelve months was too far away to shape day-to-day decisions. One month was too short to build anything meaningful without becoming reactive. Ninety days gave me enough runway to install routines, see real progress, and adjust course without losing momentum.

It also created a natural cadence. Four sprints a year. Clear starts and finishes. Regular chances to review what worked, what didn't, and what needed to change, without pretending the original plan was perfect.

Most importantly, it gave me a way to move the department even when the wider machine kept doing what large organisations do. The sprint created a protected lane for progress inside a system that otherwise defaults to maintenance.

I didn't think of this as a framework at the time, but in hindsight it mirrors the idea of habit stacking. You don't change everything at once. You layer behaviours until the system starts carrying itself.

When I stepped into H&C, the first ninety days weren't about big outcomes. They were about habits. Values conversations. One-on-ones. Clear standards. Predictable rhythms. Expectations that didn't change week to week. None of that was exciting, but it gave the department something solid to stand on.

The second ninety days built on that base. Execution tightened. Capability improved. We started shaping the department toward what was coming next. That mattered, because the third sprint was always going to look different. Exercise Talisman Sabre was on the horizon, and when it arrived, most of our focus had to shift to sustained delivery.

That flexibility is what people often miss with department goals. The structure stays consistent. The emphasis changes.

Some sprints were about installing habits. Others were about lifting capability. Others were about execution during high-demand periods. The goals changed shape based on what the department needed at that time.

Early sprints were simple. Disciplined handovers. Consistent routines. Training foundations. Clear expectations. Nothing heroic. Just stability.

Later sprints added another layer. Then another.

That's how capability compounds. You don't jump from chaos to elite in a single quarter. You move from chaos to stable, stable to sharp, and sharp to competitive. One sprint at a time.

In my case, I was fortunate. By the time department goals came into play, I wasn't dragging a broken system into shape. I was moving a sharp department toward competitive, and then competitive toward elite.

The ninety-day sprint gave me a way to do that deliberately, without breaking the people or relying on constant pressure to force momentum.

What I did in Hospitality and Catering

When I took over H&C, I made a clear call early. We were going to build a department that could legitimately compete for the Silver Platter. Not as an idea or an aspiration, but as a standard we were prepared to be judged against.

That decision created immediate clarity.

Departments don't end up competitive by accident. To even be in that conversation, several things have to be true at the same time and hold consistently. Hygiene has to be tight. Service has to be sharp. Standards can't depend on who's on shift. Training has to be deliberate. Leadership has to be visible. Pride has to exist at the deckplate level, not just in briefs.

Once that standard was set, the work became a reverse-engineering exercise.

If the department was going to be assessed as one of the best, presentation had to be world-class every day, not just during inspections or VIP visits. Customer expectations had to be exceeded as a baseline. Consistency had to be boringly reliable.

That framed the next step clearly. What do we need to build in the next ninety days to move in that direction?

Rather than trying to fix everything at once, I set a small number of goals tied directly to the levers I could pull. People. Training. Delivery. No more than three goals per lever in any sprint. Each goal had a named owner and a clear outcome.

I kept the goals simple because simplicity scales.

Every goal had to pass one test. At the end of ninety days, would we know whether it had been achieved or not? If the answer required explanation or qualification, the goal wasn't ready.

Across the sprints that worked, the structure was consistent. A defined outcome, a way to verify it, a broad approach, and a fixed ninety-day window. Anything beyond that added complexity without improving execution.

I was strict about volume. Three goals per sprint, sometimes four, never more than five. Anything more than that wasn't prioritisation. It was dilution. Limiting goals forced real trade-offs and gave the team permission to deprioritise work that didn't align with the sprint.

Talisman Sabre is where this approach was properly tested.

Going into that period, we had very little reliable information. Communication from headquarters was limited. Numbers shifted constantly. Estimates ranged from an additional 350 people per meal up to 750, on top of our existing demand.

For context, the galley was designed to feed around 80 people. We were already feeding 250 to 300 HMAS *Coonawarra* and US Marine personnel, and we were doing it at roughly 60 percent staffing.

For that sprint, the primary goal was set in simple, operational terms.

Over the ninety-day period covering Exercise Talisman Sabre, the department will feed all assigned personnel safely, consistently, and on time, while maintaining core standards and avoiding team burnout.

That was the outcome.

Under that, we had supporting goals tied to delivery and people. Shift structure. Fatigue management. Output continuity. None of them competed with the main objective.

I took the team out for breakfast, about twenty sailors, and laid it out. This is the situation. This is the goal. It's going to be heavy. Now let's work out how we do it without breaking.

The solution didn't come from me. The team proposed running a 24/7 galley. Staggered shifts. Continuous output. Dropping the idea of normal hours in an abnormal environment.

That adjustment changed everything. We met the demand, supported the exercise properly, and avoided burning the team trying to force a peacetime model onto an operational problem.

It worked because the goal was narrow and explicit. Everyone knew what did and didn't matter for that sprint. Decisions could be made quickly without escalating everything back up the chain.

That experience reinforced something I'd already learned. Department goals are only useful if they survive messy weeks. They need to be clear enough so that people can make good decisions when information is incomplete and conditions are changing.

That's why I never treated goal-setting as a theoretical exercise. It's a practical management tool. A small number of goals. Clear outcomes. A defined window. Enough structure to guide decisions without collapsing the moment reality intervenes.

That's how department goals earn their place in the Operator System.

Ownership Still Matters Here

Department goals sit after accountability charts within the Operator System for a reason. A goal without ownership stays theoretical. When a goal has a single name next to it, it stops being an idea and starts being carried. Attention has a clear place to land, progress has a reference point, and the work stops floating between "someone should" and "we'll get to it".

Giving a goal an owner does not mean that person does all the work. It means they carry the outcome. They track progress against the goal, make movement visible, and keep the score. Delegation still happens. Support is expected. Accountability remains anchored. The difference is that responsibility for the result is clear and stable across the sprint.

When I ran H&C, this distinction mattered. Goals with owners behaved differently. Updates were specific instead of vague. Decisions tightened because someone was accountable for the outcome, not just the activity. Conversations became simpler and more productive because they went straight to the person carrying the goal, with a shared understanding of what progress looked like.

This is the management function of ownership at this level. With an owner in place, goals can be tracked without chasing, reviewed without guessing, and discussed without theatre. They start influencing daily decisions about what work gets prioritised, what gets deferred, and what gets pushed back on. At that point, the question is no longer whether the department can operate. It's whether it's moving.

Weekly Rhythm

Most department goals I'd seen didn't fail because they were poorly written. They faded after the kickoff.

Week one was usually solid. The goals were clear, people were engaged, and everything looked good on the board. Then normal conditions returned. Service tempo increased. Staffing shifted. Inspections appeared. Someone posted out. An incident pulled attention. By the middle of the quarter, the goals still technically existed, but no one could point to what had changed since the start.

That gap sat with me.

I didn't close it with more meetings or reporting. I closed it by building a rhythm that stayed consistent regardless of what the week exposed.

Once a week, I checked sprint progress with the goal owners. Short, direct, and non-negotiable. No slide decks. No long forums. Just a disciplined review that stayed intact even when everything else was competing for attention.

For each goal, we called the status: on track, at risk, or off track. Then we answered three questions. What moved this week? What's stuck? What's the next action?

That was the entire process.

The point wasn't reporting up the chain or generating paperwork. It was visibility. If something drifted, it showed up early. If something was blocked, it surfaced before it became a problem. If priorities needed to shift, we adjusted them instead of letting them slide unnoticed.

Over time, what became clear was that goals didn't disappear because they were wrong. They disappeared when they stopped being looked at. This rhythm kept them present week after week, even when attention was being pulled in every direction. Once that cadence was in place, goals stopped feeling like a quarterly exercise and started behaving like part of the system. They stayed active because someone was checking progress regularly, asking the right questions, and making small adjustments before things compounded.

Where This Takes Us Next

Once department goals are in place, the department has direction. What changes next is the manager's role.

Up to this point, the Operator System has been about stabilising the environment and aiming for collective effort. The department knows where it's going. Priorities are set. Ownership is clear. The work is moving.

The next layer is about how the manager supports people inside that system. Individual goals are not separate from the work, and they're not something people chase on top of their jobs. They're how the manager helps people grow through the work that already exists. Done properly, they don't compete with department goals. They reinforce them.

This is where development stops being aspirational or optional and becomes part of how the system operates day to day. The manager's job shifts from setting direction to supporting progress at the individual level without breaking momentum or adding load. That's what the next chapter is about. How I used individual goals inside the Operator System to support people's growth while the department continued to move.

OPERATOR SYSTEM

THE MASTERY
Lead like a pro

THE EXECUTION
Build the rhythm

Level 9
Daily Systems = Self Management

Level 8
Battle Rhythms = Management

Level 7
One-On-One Meetings = Management

THE STRUCTURE
Build the how

Level 6
Individual Goal Setting = Leadership

Level 5
Department Goals = Management

Level 4
Accountability Charts = Leadership

THE FOUNDATION
Build the why

Level 3
Processes & Procedures = Management

Level 2
Standards & Expectations = Leadership

Level 1
Vision, Mission & Values = Leadership

Twenty-One

Individual Goal Setting — Developing Your People

A leader is best when people barely know they exist; when the work is done, the aim fulfilled, they will say: we did it ourselves.   — Lao Tzu

Individual goal setting matters to me more than most leadership tools because it's the moment where the mask comes off. It's the point where you stop dealing with job titles, performance metrics, rosters, and task lists, and you see how a person is building their life while they're turning up to work for you.

You're no longer listening to what gets said in a stand-up or reading what sits neatly in a position description. You're seeing what they're aiming at and where they believe their future is headed.

As a manager, this is one of the rare moments where you're allowed inside someone's long game. If you rush it, treat it lightly, or reduce it to a tick-box exercise, you waste one of the few chances you get to lead someone at a level that matters.

Professionally, Defence gets some things right, and formal individual goal setting is one of them. Annual and mid-year performance reviews — SPARs — cop plenty of criticism, but I never hated them. I looked forward to them. Feedback mattered to me then, and it does now. I wanted to know where I stood, what I was doing well, and where I was falling short. I didn't need it to be softened. I needed it to be honest.

That perspective shifted the moment I started writing them myself. It didn't take long to realise how difficult it is to articulate someone's performance properly, especially when you're trying to be fair, accurate, and useful at the same time. When I first started writing performance reports, I was terrible at it. There's no polite way to dress that up. Chief Petty Officer Lyndon Quirke could attest to that without hesitation. I overthought every sentence, second-guessed every word, rewrote paragraphs repeatedly, and still walked away feeling like I'd missed something important.

Even now, writing this book, that same doubt shows up. I still catch myself wondering who I am to write about leadership at all. That voice never disappears, but you learn to recognise it for what it is and refuse to let it make decisions for you.

What I'm talking about here, though, isn't professional goal setting. Defence has that space well covered. Where I've seen things start to break down is when leaders stop at performance management and never step into personal direction. I've watched it happen repeatedly, both in uniform and outside it, and the cost is always the same: people turn up, do the work, and still feel disconnected from where their own life is heading.

What This Conversation Is Really For

Personal goal setting is about understanding where your people want to be in two to five years, not just inside the organisation, but as human beings with lives that extend well beyond their role description.

When I sit down with someone, I'm trying to understand what they're aiming at across a few simple areas. Not in a polished or rehearsed way, but honestly, as they see it at that point in their life in specific areas:

- Personal

- Family

- Finance

- Health

- Wealth

- Career

- Other.

That last category matters more than most people expect. It's where things sit that don't fit neatly into a professional framework but still shape how someone shows up every day. Faith. Belief. Experiences they want to have. Travel. Education. Interests that will never appear on a performance report but absolutely influence how a person thinks, decides, and commits.

We'll break each of these areas down properly later in the chapter, but this is the starting point. This is the moment where you stop guessing what matters to someone and give them the space to tell you, in their own words, in what direction they believe their life is heading.

This isn't counselling. It isn't therapy. And it isn't about fixing anyone.

It's about listening long enough to understand the trajectory they're aiming for and the trade-offs they're willing to make to get there. Once you understand that, the work in front of them stops feeling random. It starts to make sense in the context of something bigger than the task list.

That's why Individual Goal Setting sits so high in the Operator System. Every layer beneath it — standards, clarity, accountability, and capability — only truly holds when it connects to something the individual cares about. Without that connection, you're managing behaviour. You're not developing people.

When you understand someone's direction, leadership decisions change. If a sailor is aiming for financial stability so they can buy a house in three years, you start looking for options that support that. If someone wants to move into a different career stream, you stop treating it as disloyalty and start helping them build a realistic pathway. If someone's health is slipping because they're grinding themselves into the ground, stepping in early becomes part of your responsibility.

This is where leadership stopped being theoretical for me. Not because I'd read the right book, but because I'd seen what happened when this step was ignored and what changed when it wasn't.

When people can see how their work connects to where they're trying to go, effort changes shape. Responsibility gets picked up without being forced. Standards lift without constant pressure. Decisions improve because the work no longer feels disconnected from their own life.

That shift didn't come from saying the right things. It came from listening first, then being deliberate about how work was assigned, developed, and supported afterwards.

In practice, this has reduced how much pushing I've ever needed to do as a leader. When people understand why the work matters to them, not just to the organisation, ownership shows up without being chased.

That's why this layer sits where it does in the Operator System. It's the point where structure stops being abstract and starts becoming human. Standards and systems don't disappear, but they make sense in the context of someone's life, not just their role.

If you're willing to sit across from someone, listen, and take their direction seriously, the relationship changes. People stop feeling like they work *for* you and start experiencing the work as something you're doing *alongside* them.

That's what development has looked like in practice for me. Not performative. Not theoretical. Just deliberate choices made with a real human in mind.

Why Individual Goals Sit *After* Department Goals

Individual goal setting sits here in the Operator System because development only works when direction already exists.

I didn't design it that way on paper. I learned it by watching what happened when I got the order wrong. When I tried to develop people without first being clear on where the department was heading, effort scattered. People grew in different directions, pulled toward different futures, and the team lost cohesion without intention.

Department goals answer the shared question first: where are we going? Individual goals answer the next one: how do you grow while we're moving there?

Once that sequence was clear, individual development stopped competing with the work and started reinforcing it. Skills were built with purpose. Responsibility was taken on with context. Growth felt useful instead of abstract, because people could see how it served both the department and their own direction.

That's the benefit of this layer's position. It turns development into something practical, aligned, and sustainable, rather than well-intentioned noise. When direction comes first, individual goals don't fragment the system. They strengthen it.

What Individual Goal Setting Actually Is (and Isn't)

This isn't HR administration, and it was never meant to be treated like another process to manage. It isn't a form you complete, file away, and revisit twelve months later because the calendar tells you to. And it isn't about promising promotions, postings, or outcomes you don't control and were never in a position to guarantee.

When I did this properly, individual goal setting wasn't a process at all. It was leadership intelligence.

It was how I learned what actually drove the people sitting across from me, not what they thought they were supposed to say. It gave me visibility into what they were aiming for, what they were worried about, and what they had already written off as unrealistic because no one before me had taken the time to ask. Patterns emerged quickly. Who was carrying financial pressure? Who was running low on physical or mental capacity? Who was ambitious but unsure how to articulate it? Who had already decided growth wasn't for them, even though they still showed up and did the work?

That insight altered how I led, whether I intended it to or not.

Once you understand those things, assumptions fall away. Behaviour that once looked like resistance reads differently. Hesitation has context. Quiet compliance stops being mistaken for contentment. You can allocate work deliberately instead of reactively, and conversations become more direct because you understand what is at stake for the person in front of you.

Without that insight, you're guessing. You're interpreting behaviour without context and making decisions with partial information. I've done that myself, and I've watched others do it around me. More often than not, those guesses miss the mark, and the consequences land with the person who has the least room to absorb them.

Individual goal setting, done properly, isn't about managing outcomes. It's about understanding people well enough to lead without relying on assumption, habit, or hope to fill the gaps.

The Areas I Always Covered

I never ran these conversations like an interview, and I never treated them like a checklist. I wasn't trying to extract information or fill gaps in a template. Over time, though, I learned there were certain areas you couldn't afford to ignore. Every time one of them was left untouched, it surfaced somewhere else as friction, frustration, or disengagement.

We always started with the personal side, because pretending people leave their life at the gate is dishonest. That meant talking about family pressures, relationships, kids, ageing parents, and the realities outside the uniform that still shaped how they showed up to work. It also meant talking about what they wanted their life to look like in a few years, not in abstract terms, but practically. Where they wanted to live. What stability meant to them. What they were trying to protect or build.

From there, we talked about finance and wealth. Money stress has a way of bleeding into everything if it's left unspoken. Sometimes it was debt. Sometimes it was saving. Sometimes it was simply feeling stuck without a clear way forward. I didn't need to be a financial expert, and I never tried to be. What mattered was understanding whether money was a background consideration or a constant weight, because that shaped how people approached risk, opportunity, and responsibility.

Health was always part of the conversation as well, even when it was uncomfortable. Physical capacity, mental load, fatigue, and recovery all mattered. In high-tempo environments, pretending those things don't exist doesn't make you resilient; it just makes you blind. You can usually tell when someone is running close to empty, but unless there's space to speak about it, people keep pushing until something gives.

I'd already learned that lesson the hard way earlier in my career, which I've covered in previous chapters. By the time I was having these conversations with others, I didn't need theory to convince me. I'd lived the consequences of ignoring health signals and mistaking endurance for sustainability.

That experience reshaped how I look at people, particularly in leadership and hiring. At Neptune, this became a practical consideration, not an ideological one. I'm not interested in aesthetics or performative standards. What I care about is whether someone can sustain performance in a fast-paced, high-pressure environment. Fitness, recovery, and self-management aren't about image; they're about having the capacity to think clearly, handle pressure, and keep showing up without burning out. When those fundamentals are neglected, performance degrades, no matter how capable or driven the person is.

Career came next, and this was where conversations often surprised people. We didn't just talk about promotion or the next rank. We talked about what kind of responsibility they wanted, what work gave them energy, and whether leadership was something they felt drawn toward or something they were quietly avoiding. Not everyone wants the same path, and treating them as if they should only creates resentment on both sides.

We also talked about capability. What they wanted to be better at. Where they felt exposed. What skills they knew they were missing but hadn't had the chance or confidence to admit. Those gaps weren't problems to be fixed. They were signals that helped me understand where development would matter.

There was always space for the "other" category. Sometimes that meant study. Sometimes business ideas. Sometimes travel, faith, or personal projects that had nothing to do with Navy or Neptune. Those conversations often mattered more than they appeared to, because they revealed what gave someone energy when work didn't.

None of this required me to solve everything in the moment. But it also wasn't passive. I gave advice when it was earned, grounded in experience rather than theory. What mattered most was listening without rushing to an outcome, and then acting on what I'd heard.

Time Horizons That Made Sense

I never asked people to map out their entire life in one sitting. I tried that early on and learned quickly that it overwhelms most people. When you push too far, too fast, you don't get honesty. You get safe answers that sound reasonable and don't mean much.

What worked better was using time horizons people could imagine.

We started with the next twelve months. This wasn't about reinvention. It was about momentum and confidence. Short-term goals gave people something tangible to aim at and something they could realistically complete. A course finished. A qualification started. A capability strengthened. Sometimes it was as simple as clearing something they'd been putting off because no one had ever made space for it. Early progress mattered, because visible progress changes how people show up.

Then we looked further out, usually five years. Not as a promise or a plan, but as a direction. This was where I asked people to slow down and think about what success would look like if things went well. What their days might look like. What responsibility they'd be carrying. What kind of operator they were to others. I wasn't interested in locking anyone into a path. I was interested in understanding how seriously they were thinking about staying, growing, or transitioning.

That five-year view told me a lot. Not because it was accurate, but because it revealed intent.

Alongside that, I added one exercise that consistently shifted the conversation. I asked people to imagine money was no longer a constraint and to write down one hundred things they wanted to do in their life. Places they wanted to go. Experiences they wanted to have. Study, family goals, work goals, indulgent ideas, unrealistic ideas. Anything that came to mind.

I did this exercise myself, and the list surprised me. Play golf at St Andrews. Travel the UK playing the top fifty courses. Buy a Porsche

GT3 RS. Holiday in the Swiss Alps. Watch Manchester United at Old Trafford. Retire from the Navy. Write a book. Seeing those written down gave depth to what actually mattered to me in a way no performance conversation ever had.

Once that list existed, the earlier goals made more sense. If something could realistically be ticked off in the next twelve months, we looked for a way to make it happen. Not because it was indulgent, but because progress that's felt is more powerful than progress that's only measured. It reinforced the idea that effort could translate into experience, not just output.

The visualisation piece mattered here. I've spent time around Johnnie Cass from Entourage, and he's exceptionally good at helping people step into what achievement feels like rather than just talking about it. I don't operate at that level, but the principle holds. When someone can picture themselves having achieved something, the goal stops being abstract. It becomes real enough to pull them forward.

None of this was about precision. Goals change. Life changes. Circumstances shift. What mattered was intent and direction, and giving people a way to connect the work they were doing now to a future they could imagine themselves living.

That's the point where these conversations stopped being administrative and started becoming useful.

How I Ran the Sessions

These conversations were never rushed or squeezed between tasks. If I was going to do this properly, they needed real space and real attention. I scheduled one a day only. They take bandwidth, and you can't half-listen and expect anything useful to come out of it.

I always started with my immediate staff. That wasn't about control, it was about leverage. Once they understood how these conversations worked, I taught them how to run the same sessions with their own people. That scaled far better than me trying to sit down with everyone myself, and it embedded the practice instead of turning it into a one-off leadership gesture.

Wherever possible, I took these conversations off-site. A café. Lunch somewhere quiet. Somewhere away from the office and away from the eyes and ears of everyone else. Phones went off completely. A different environment drops the guard people wear at work. In uniformed environments, I stripped rank out of the room. Civilian clothes. First names. No performance. Just two people talking honestly.

Once the conversation started, my job was simple. I shut up and listened.

This wasn't about fixing anyone's life or solving problems on the spot. It wasn't about rescue. My role was to facilitate, challenge, and help people land on goals they owned. If I set the goals for them, they wouldn't carry them. And if they didn't carry them, nothing would change.

I kept the structure light on purpose. Complexity kills honesty.

We talked about why their development mattered to me. We talked about what they wanted from their life and career. We talked about where they felt stuck and what they'd been avoiding. We talked about the gap between where they were and where they wanted to be. We talked about what was realistic to work on together. We talked about how we'd check back in so decisions didn't quietly evaporate.

Tracking mattered. Visibility mattered. If goals disappear into a notebook or a folder, they die. I learned a lot from people like Dan Martell. He keeps his goals as his phone screensaver so they're always in front of him. That idea stuck. Goals need to be seen, not stored. How you surface them will differ, but if they're not visible, they won't survive day-to-day pressure.

This was also where I started leaning harder into tools, especially AI. Whiteboards, sketches, diagrams, GPTs. Not to replace the conversation, but to deepen it. Used properly, these tools help ask better questions, explore trade-offs, and pressure-test thinking with people, not instead of them. They don't replace leadership. They sharpen it.

I'm blunt about this. Organisations, including Navy, need to stop shielding people from AI out of fear. The productivity gains are obvious. Teaching people how to think with tools is part of development, not a threat to authority.

That was the whole approach. No promises I couldn't keep. No false certainty. No performance theatre. What people got instead was clarity, honesty, and consistent follow-through.

A Real Example: Turning Intent into Movement

When I did this work with Joel, one of my team leaders at Neptune, the picture became clear very early. What he wanted had nothing to do with climbing a management ladder or building a long-term career in our business. He wanted to drive dump trucks in the mines.

As a kid, he'd been obsessed with trucks. Life, bills, and responsibility nudged him onto a different path, and over time that ambition got parked far enough to stop being spoken about at all. Not because it disappeared, but because no one had ever asked.

I treated that information for what it was. Data, not a threat.

We built a development plan for the next twelve months that allowed him to keep leading his team properly while preparing for where he wanted to go. I paid for his training and courses. In return, I asked for one thing: that he commit fully to his role for the next twelve months and continue showing up the way he always had.

What happened next surprised people who believe development only works when it's tied neatly to succession planning. Joel didn't give me twelve months. He gave me three years.

Those years weren't negotiated or extracted. They were offered. He stayed because he felt seen, backed, and treated like a human being with agency, not a resource being managed. When the time came, he left on good terms and went off to live the life he'd described in that first conversation.

I've always come back to the line often attributed to Richard Branson:

Train your people well enough so they can leave, and treat them well enough so they don't.

That idea isn't soft or generous. It's practical.

Joel's goal didn't align neatly with a structure chart or a long-term business plan. What it did was secure a capable team leader and a good human being for three solid years, at a point where stability mattered. That outcome didn't come from contracts, retention strategies, or pressure. It came from respecting someone's direction and being honest about what we could support together.

That's what individual goal setting looks like when it works.

Not everyone stays forever, and they shouldn't have to. The real measure is whether the time they do stay is committed, productive, and honest, rather than spent disengaged while waiting for an exit.

That's a trade I'll make every time.

Development Doesn't Require Budget

This is where I've seen a lot of leaders stop themselves long before they need to. I watched it happen repeatedly at Neptune. Managers would identify a real gap, emergency management, customer service, decision-making under pressure, and then stall. No budget. No approvals. No external course. So nothing happened.

The problem wasn't resources. It was intent.

When development is tied to a clear goal, either an individual goal or a department outcome, the question stops being "what can we afford?" and becomes "what do we need to build next?" Once you're clear on that, the excuses shrink fast.

Personal development matters to me because it's how I grew, and it's how I've seen others grow when someone invested time instead of waiting for permission. Today, access to learning isn't the constraint people pretend it is. YouTube, podcasts, short-form content, books, case studies, lived experience. You can take a twenty-second clip, pull out one useful idea, and turn it into an hour of meaningful discussion if you know what capability you're trying to build.

That's the link to goals.

If someone's individual goal is to step into leadership, then decision-making, communication, and accountability become development priorities. If a department goal requires better delivery under pressure, then emergency management, prioritisation, and coordination become development priorities. You don't need a course catalogue for that. You need clarity on what success requires next.

AI has only compressed that gap further. If there's a real issue in your team, you can describe the environment, constraints, and outcome you're trying to achieve and have a practical training session designed in minutes. Not theory. Something that fits how your people work. That doesn't require budget approval. It requires you to think about the goal you're supporting.

One of the most useful things I've learned is how learning sticks. Passive intake doesn't work well. People listen, nod, write notes, and forget most of it. What does work is teaching. When someone has to explain an idea, structure it, and answer questions, understanding deepens fast.

That's why I pushed people to teach.

If someone learned something new that aligned with their development goal or the department's direction, they ran a short session for the team. That did two things. It reinforced their learning, and lifted capability across the group without creating dependency on me or external training.

Development doesn't require budget. It requires direction.

When goals are clear, development becomes obvious. When goals are vague, leaders hide behind constraints. The difference isn't resourcing. It's whether you're willing to connect growth to outcomes and act on it.

When Goals and Performance Don't Match

Sometimes someone wants something they're not ready for. I've seen this often enough to stop being surprised by it. In some cases, people never get clear on what they want beyond the title or the status attached to it. In other cases, they're clear on the outcome, but not willing to commit to the process required to earn it.

I had an employee like this at Neptune. I liked him. I could see potential there if he ever decided to apply himself. I gave him more than enough opportunity to do that. Access wasn't the issue. Support wasn't the issue. The gap sat between intent and effort.

He believed he deserved the outcome without committing to the work.

He wanted to become a K9 handler and be part of our Public Order Response Unit within Neptune's Special Operations Division. That role isn't aspirational. It's earned. It's typically held by high-performing operators who've already demonstrated discipline, consistency, and reliability under pressure. He hadn't done that work, but he wanted the position anyway.

We gave him the opportunity.

He completed the course, which is only the entry point. Anyone who's worked seriously with K9s knows the real work happens afterward. Kennel duties. Feeding. Cleaning. Daily training. Maintaining fitness. Building trust with the animal. Showing up when no one is watching. That's the job long before anyone sees the capability.

He turned up once after the course and then disappeared. No follow-through. No ownership. No sustained effort beyond being given the opportunity. When the outcome didn't materialise, the story shifted. He claimed he hadn't been given a fair shot.

This is where I draw a hard line as a leader.

My responsibility is to create opportunity. It is not to manufacture commitment.

There's a lesson in this that I also need to own. In hindsight, I backed that goal too early. At the time, I wanted to see whether a serious opportunity would trigger the level of commitment that had been missing. That decision came from optimism rather than evidence.

It sharpened my judgement.

I learned to distinguish more clearly between ambition and readiness, and between people who want an outcome and people who are prepared to earn it. That judgement doesn't come from avoiding risk. It comes from paying attention to what happens when people are tested against the standard they're asking for.

Backing someone's goal doesn't mean saying yes to it immediately. It means assessing whether their current behaviour supports the direction they're asking for. Some goals deserve investment. Others require a period of proof before support is extended.

The point isn't to shut people down. It's to be honest about the gap. What excellence looks like. What needs to change. What support exists if they choose to step up.

Sometimes that conversation leads to growth. Sometimes it leads to the realisation that their path sits elsewhere. Both outcomes are valid.

What isn't valid is false hope.

Clarity, even when it's uncomfortable, respects people far more than pretending commitment exists when it doesn't.

Why This Layer Multiplies Everything Else

This layer sits where it does because, when it's done properly, it changes how the rest of the system behaves. I didn't arrive at that conclusion through theory. I learned it by watching what happened when individual development was treated as part of the work instead of something added on when time allowed.

When people are developed with intent, responsibility starts shifting naturally. They stop waiting to be managed through every decision and begin exercising judgement for themselves. Not because expectations were lowered, but because they were trusted with work that stretched them and held them to a clear standard. Over time, the strongest performers don't just lift their own output. They stabilise the environment around them.

That's when departments stop being fragile.

You're no longer relying on a small number of people to hold everything together. Capability spreads. Decisions improve closer to the work. Pressure gets absorbed instead of amplified because more people can carry it without escalation.

Department goals provide direction, but direction alone doesn't create momentum. Individual goals are what convert that direction into sustained movement. They allow people to see how their own growth connects to where the team is heading and what's required to get there. When that connection is clear, effort doesn't need to be forced or constantly reinforced.

This is also where leadership changes shape. The workload doesn't disappear, but it stops being exhausting in the same way. You're no longer compensating for disengagement or dragging people toward outcomes they don't recognise as meaningful. You're working with people who understand the direction and have chosen to invest in it.

Everything that follows in the Operator System assumes this foundation layer exists. Without it, the system can still function. Standards can hold. Output can continue. But it never really scales, and it never becomes resilient.

With it, progress compounds.

OPERATOR SYSTEM

Twenty-Two

One-on-One Meetings — Your Leadership Rhythm

Discipline is the bridge between goals and accomplishment.   — Jim Rohn

Entering the Execution Phase: Building the Rhythm

Early in my management career, long before I had any formal system to follow, execution kept slipping in ways that pissed me off because I couldn't pin down why. Direction was clear. Work had been handed out. People knew what they owned. Yet every week felt the same. The same issues repeated. Deadlines crept. I spent most of my time responding instead of driving anything forward.

What I was missing wasn't competence or intent. It was cadence.

Most of my day disappeared into unplanned one-on-one conversations. Quick check-ins that were meant to take five minutes and blew out to thirty, especially with my business partner, because fuck me, Bondi can talk. People grabbed me between meetings, dropped questions on me mid-task, or unloaded problems whenever they hit a wall because I'd never programmed a proper time for it. I told myself I was being accessible. In reality, I was bleeding time and attention across the day.

I was constantly switching context, leaving conversations without clear outcomes, getting pulled into the next thing, then trying to remember what I needed to follow up later. Multitasking felt productive in the moment, but it fractured my day and tore execution apart. I was busy from morning to night and still getting fuck all done, and that wasn't bad luck. It was self-inflicted.

The turning point came when Shane Byrne at the Entourage introduced me to the concept of structured one-on-ones. What landed wasn't the meeting itself, but the discipline behind it. Those individual conversations finally had a programmed place to live. Once they moved into a fixed weekly cadence, execution started to stabilise. Priorities held. My time stopped getting hijacked by whoever caught me next.

That cadence started with one-on-one meetings.

One-on-Ones: Where Rhythm Actually Starts

I've never cared whether one-on-ones were popular. I cared whether execution held together. Most of the resistance I've seen comes from using them for the wrong job. When one-on-ones are treated as a place to decide direction, they feel slow and inefficient because the conversation keeps spilling elsewhere.

That's not what they're for.

Direction isn't set in one-on-ones. It's already been set. The role of the meeting is to turn agreed direction into output at the individual level, week after week, without renegotiation. Once you stop asking them to carry decisions they're not designed for, the meeting becomes clean and useful.

The structure I landed on was simple and disciplined because it had to survive real weeks, not ideal ones.

Each one-on-one starts with a brief check-in. A few minutes to acknowledge the human across the table, then we get to work. Left unmanaged, that space can drift into venting or therapy, so it's deliberately bounded. The point isn't comfort. It's containment.

From there, we look at the previous week. What landed, what didn't, and how outcomes compared to what was expected. This works best when there's something concrete to look at, whether that's a CRM, notes, or a task list, but it doesn't need to be sophisticated. I'm not looking for updates. I'm looking for patterns. Sometimes that shows progress I wasn't aware of. Other times it flags slippage while there's still room to adjust. If something is off, that's where I bring the S.O.F.A. model into the conversation or deal with an issue that irritated me during the week but didn't warrant stopping everything in the moment.

When something hasn't worked, we deal with it. The focus is practical. How do we stop this repeating, and what do you need from me to make that happen? That keeps the conversation anchored on follow-through instead of revisiting the past.

We then move into the week ahead, and this is where calendars earn their keep. I want to see what the week looks like, not hear a verbal plan. Sales calls, site visits, training blocks, and other real commitments should already be locked in. A calendar shows output, pressure points, and spare capacity faster than any update ever will. I live by a simple rule. If it isn't in the calendar, it doesn't happen.

Once this cadence was in place, individual-level signal stopped getting lost. I could see pressure building before it turned into behaviour. I could see capacity issues forming before deadlines slipped. Patterns across people became visible instead of every issue landing as a surprise. Conversations stopped spilling across the week and started landing where they could be handled.

That's where rhythm started doing real work. Individual execution stabilised, my time stopped leaking across the day, and momentum stopped depending on interruption. One-on-ones became the mechanism that kept execution contained, predictable, and moving.

What Changed When That Rhythm Held

What one-on-ones solved for me wasn't big failures. It was the loss of an early signal at the individual level. Without a disciplined cadence, that signal thins out one person at a time. Hesitation goes unnoticed. Load gets carried silently. Strong performers flatten under sustained pressure that isn't visible anywhere else.

With a fixed one-on-one rhythm, that signal stayed intact. I wasn't reacting to drift after it had taken hold. I could adjust workload, expectations, or support while execution was still moving.

That only works when one-on-ones stay disciplined and consistent.

They're not where direction gets renegotiated. They don't replace team forums or decision bodies. They aren't a place to unload unresolved issues or ambush performance. When they're asked to carry those roles, they lose their function.

Used properly, one-on-ones provide execution containment at the individual level. That containment is what allows the rest of the execution phase to operate without constant intervention.

The Structure That Works

Early on, I overcomplicated one-on-ones and paid for it. I made them too long, tried to cover too much, and let agendas wander. Meetings drifted because I treated them like open conversations instead of a controlled execution tool. The signal that sent was simple. This wasn't important, and nothing kept its shape.

What worked had to survive reality, so I stripped it back.

I ran one-on-ones weekly and kept them short. Fifteen to twenty minutes was enough when the structure was right. They were private, distraction-free, and always had an agenda built on the exact formula

I laid out earlier. Without that structure, the conversation defaulted to whatever felt loudest in the moment instead of what mattered.

When I introduced one-on-ones into the H&C team, I set the agenda at the start. I wanted to model the standard before handing it over. After about four weeks, ownership shifted to them because it was their meeting with me, not my meeting with them. That shift mattered. Preparation improved. Accountability sharpened. The conversation became something they used instead of something that happened to them. It also gave me a far clearer view of what mattered to them, not just what I thought should matter.

Once individual goals were set, I added them to the bottom of the agenda and kept them there. That was a hard break from how I used to operate, and from how I still see too many senior sailors operate. Set goals. Forget them. Drag them out at the six-month mark. Dust them off again at reporting time. That's not managing people. That's lying to yourself about doing the work. You cannot shape execution if you look at someone's goals twice a year. If they matter, they need to be visible every single week. No fucking exceptions.

That's where evidence came in.

I wasn't interested in explanations or intentions. I wanted to see what had moved since last week. What had been done. What had changed. Evidence kept the conversation grounded and stopped one-on-ones from turning into opinion or optimism. Execution either showed up or it didn't.

Every meeting finished the same way: with one question. What do you need from me? Sometimes the answer was practical. Sometimes it was political. Sometimes it was developmental. What mattered was that the question was always there. Over time, people learned it wasn't a courtesy or a throwaway line. It was a real handover point, and when something was raised, it got acted on.

That closing mattered more than most people realise. It stopped issues being carried silently. It kept accountability two-way. It also made it clear that execution wasn't just something I measured; it was something I supported.

Keeping the meetings short forced focus. The weekly cadence built reliability. A visible agenda kept everything anchored in reality. We didn't try to cover everything. We covered what mattered that week and trusted the rhythm to carry the rest.

Making One-on-Ones Survive Busy Weeks

The fastest way one-on-ones died for me was when I cancelled them as soon as shit got busy. It was the most destructive thing I did to the rhythm of my teams and to the culture early on, and I made that mistake more than once.

Every time I did it, the pattern was the same. At Neptune, it hit hardest. Small issues that should have been handled early stacked up because there was nowhere for them to land. Conversations that were meant to drive execution collapsed into surface-level updates, because no one had a proper forum to think, challenge, or reset priorities. The meetings that survived drifted toward what I wanted reported upward instead of what needed to move on the ground.

Letting the cadence slide changed behaviour fast. Cancelling one-on-ones taught people exactly how optional they were. Commitments started slipping because no one expected them to be revisited. People stopped raising things early because experience told them nothing would happen until it was already a problem. When I eventually had to address performance, it landed heavy and confrontational, not because the issue was big, but because I'd allowed it to build unchecked.

When we finally sat down again, the damage was done. Too much had gone unsaid. Frustration was baked in on both sides. Conversations took more effort than they should have, and I was cleaning up issues I'd created by dropping discipline when pressure increased.

That experience locked in a rule for me. Silence costs more than short conversations. Ten focused minutes beats forty minutes of clean-up later, every time. Cancelling a one-on-one under pressure doesn't save time. It just pushes the cost forward and makes it worse.

That's why one-on-ones are non-negotiable inside the Operator System. Execution doesn't survive without cadence. When things get busy, I adjust the meeting. I shorten it. I tighten the agenda. I do not remove the rhythm. The moment that cadence disappears, problems start compounding immediately.

That rule exists because busy weeks are the test. They're where execution discipline either survives or collapses, and everything downstream follows from that.

How This Layer Connects Upward

One-on-ones are where execution becomes visible at the individual level. They're where priorities get reset, pressure gets checked, and work either moves or gets corrected before it drifts. If something is off with a person's execution, this is where it shows up first.

When one-on-ones run properly, daily work becomes easier to manage. Priorities are clearer. Follow-ups don't rely on memory. Tasks don't stall or bounce around because responsibility and timing are being checked every week instead of guessed at. I don't need to be constantly available to keep things moving, because the cadence is already doing that job.

As that rhythm settles in, decisions improve closer to the work. Escalations arrive with context instead of frustration. I stop being the centre point propping everything up through availability and reaction, and the system starts carrying more of the load on its own.

That's the real role of one-on-ones. They stabilise individual execution so it can be coordinated at scale. Without that base, everything above becomes reactive. With it, execution can be organised instead of chased.

One-on-ones feed directly into your battle rhythm as a manager. What you see and hear in those conversations is what shapes team forums, workload distribution, and decision timing. Once the individual cadence is solid, rhythm can scale beyond one person and start moving the team as a unit.

That's where we go next.

OPERATOR SYSTEM

THE MASTERY
Lead like a pro

THE EXECUTION
Build the rhythm

Level 9
Daily Systems = Self Management

Level 8
Battle Rhythms = Management

Level 7
One-On-One Meetings = Management

THE STRUCTURE
Build the how

Level 6
Individual Goal Setting = Leadership

Level 5
Department Goals = Management

Level 4
Accountability Charts = Leadership

THE FOUNDATION
Build the why

Level 3
Processes & Procedures = Management

Level 2
Standards & Expectations = Leadership

Level 1
Vision, Mission & Values = Leadership

Twenty-Three

Battle Rhythm — Your Leadership Operating System

Plans are worthless, but planning is everything. — Dwight D. Eisenhower

Battle rhythm is a term used in military operations to describe a deliberate cycle of command, staff, and unit activities that synchronise current actions with future operations and enable timely decisions across an organisation. It exists for one reason. To make sure execution keeps moving when conditions change and pressure increases.

That definition matters because it strips away the bullshit. Battle rhythm isn't about rigid schedules or control for its own sake. It's about creating a repeatable structure that allows decisions to land at the right time, with the right information, without relying on luck or constant availability. When the rhythm is clear, work flows. When it isn't, leaders compensate with energy, urgency, and personal intervention.

I started using the term battle rhythm years before the Operator System existed, first inside FLSE-D and later at Neptune, because it matched how execution felt when things were moving fast. The name didn't land with everyone. Joel King, who was my Client Services Manager at Neptune, hated it. He thought it sounded rigid, aggressive, and thought it would kill flexibility. I understood the reaction. I'd had the same concern earlier in my career.

What experience taught me was the opposite. Discipline, applied deliberately, doesn't remove freedom. It creates it. When cadence is clear, standards are set, and time is accounted for, decision-making becomes simpler. You don't waste attention deciding what matters every day because the structure already tells you. Work stops being shaped by interruption and starts being shaped by intent.

Up to this point, the Operator System has focused heavily on individuals. One-on-ones. Personal goals. Development. That layer was necessary, but it only takes you so far. Battle rhythm is where execution moves beyond individuals and starts functioning at a team, department, and organisational level.

This layer isn't about managing people minute by minute. It's about establishing a predictable cadence for communication, decision-making, and review so execution doesn't depend on heroics. When rhythm is absent, leaders end up carrying the system with availability and reaction. When rhythm is present, the system carries the load.

This chapter is about building that operating system at scale. Not daily personal habits, but the shared cadence that allows work to move cleanly across teams without constant intervention. When battle rhythm is in place, leadership stops feeling scattered and starts feeling deliberate. Execution becomes something you organise, not something you chase.

What Battle Rhythm Means (Without the Bullshit)

Once battle rhythm is established, the organisation stops improvising how it works. There is a known cadence for alignment, decision-making, review, and communication, and that cadence governs how effort gets applied across the team.

This isn't about personal time management or individual productivity. It's about how a group of people coordinate work without constant clarification, escalation, or intervention. Battle rhythm defines when priorities are reset, when issues surface, where decisions land, and how information moves through the system.

When that cadence is clear, teams stop guessing. They know when alignment happens and where to take problems instead of carrying them or raising them at random. Work no longer depends on hallway conversations, chance availability, or who happens to be under the most pressure at the time. The organisation operates on a shared tempo instead of reacting week to week.

In the Navy, this discipline is non-negotiable. If people don't know when briefs occur, when decisions are made, or how information flows between levels, execution degrades through friction. Handovers slip. Decisions lag. Rework increases. Performance suffers even when effort remains high.

Business environments behave the same way. The difference is that the degradation is often normalised instead of addressed.

Without a battle rhythm, urgency dictates behaviour. Tactical issues crowd out coordination and planning. Decisions stack up because there's no defined place for them to be processed. Leaders become bottlenecks, not by choice, but because the system gives them nowhere to hand work back. Teams burn energy reacting instead of sustaining momentum because nothing is anchored in time.

Battle rhythm corrects that by putting structure around how the team operates together. It establishes a predictable cycle for team meetings, one-on-ones, decision forums, external obligations, and review points. Execution no longer relies on leader availability to stay aligned. The system carries that load.

This structure doesn't remove flexibility. It protects it. When the rhythm is clear, non-negotiables stop being sacrificed to noise. Teams know what can move, what must wait, and how change gets processed. Decisions can be made decisively because the boundaries and timing are understood.

That's what battle rhythm does. It protects the organisation's non-negotiables and allows real work to move consistently without constant intervention, even when conditions aren't calm.

Why This Sits After One-on-Ones in the Operator System

One-on-ones were never meant to stand alone. From the start, they were designed as one input into a larger execution rhythm.

At the individual level, one-on-ones give you a clean signal. You see where people are tracking, where pressure is building, where priorities are slipping, and where effort isn't translating into output. That information matters, but only if it feeds something bigger than the conversation itself.

Battle rhythm is that next layer.

It's where the information coming out of one-on-ones gets used to shape how the team operates week to week. What I hear in those individual conversations informs how I set priorities in team forums, where I apply pressure, where I pull it back, and how I structure the week ahead. The rhythm doesn't replace one-on-ones. It absorbs them.

If I hear the same capacity issue from multiple people, that feeds directly into workload planning. If the same friction keeps appearing across different conversations, that informs process changes or decision timing. If priorities are drifting at the individual level, that shows up immediately in how the next alignment session is run. Nothing lives in isolation.

This is also where execution stops relying on memory. Instead of holding information in my head and trying to adjust on the fly, the rhythm forces those decisions into the calendar. Team meetings, reviews, and planning sessions exist because they're the places where an individual signal gets converted into coordinated action.

One-on-ones give you visibility at the individual level. Battle rhythm turns that visibility into movement across the team. Together, they create a closed loop between what's happening on the ground and how the organisation responds.

Without that connection, execution fragments. With it, coordination becomes deliberate, repeatable, and scalable.

What Battle Rhythm Changes in Practice

Once battle rhythm is in place, work stops being driven by reaction and starts being guided by cadence.

The most immediate change is predictability. The team knows when alignment happens, when decisions are made, and where issues belong. That alone removes a significant amount of friction. People stop guessing when to raise something or hovering around leadership waiting for the right moment. Conversations land where they're meant to, instead of interrupting everything else.

Rhythm also protects the work that matters. Important tasks stop being pushed aside by whatever feels loudest that day. Planning, coordination, and thinking time have a defined place in the week instead of being squeezed in between interruptions. That's what allows depth to exist alongside execution, rather than everything collapsing into surface activity.

It also changes how much mental load the system places on the leader. When priorities are embedded into the structure of the week, you're no longer re-deciding what matters every morning. The rhythm carries those decisions forward. Instead of tracking everything mentally or relying on reminders and personal discipline, the cadence itself keeps work moving and issues visible.

This is where execution becomes reliable. Not because people are more motivated, but because the system no longer depends on energy, memory, or constant follow-up to function. When pressure increases or conditions aren't ideal, the rhythm keeps the work coordinated without everything defaulting back to reaction.

That's what battle rhythm changes in practice. It doesn't make the work easier. It makes execution sustainable, so performance doesn't fall apart the moment attention gets stretched.

The Six Components I Never Compromised On

When I built battle rhythm at a team level, I wasn't trying to cover everything. I was trying to make sure the right things were always present. Over time, I landed on six components that had to exist inside the team cadence for execution to stay organised. The environment changed. The tempo changed. Those six didn't.

1. One-on-ones — were built into the rhythm because they fed everything else. They weren't a separate leadership habit running alongside the system. They were the primary input. What surfaced there shaped how I planned alignment sessions, where pressure needed to be applied, and what decisions had to be made at a team level. Without that individual signal feeding into the rhythm, the rest of the cadence ran on assumption.

2. Team alignment — where the rhythm moved from individual execution to collective coordination. This showed up as management meetings, divisional forums, or team briefings, depending on the environment. The purpose was always the same. Reset priorities, make changes visible, and ensure effort converged instead of scattering. When this wasn't built into the rhythm, teams stayed active but drifted strategically.

3. Training — earned its place because it moves departments from functional to elite faster than anything else. That's why it had to live inside the battle rhythm. If training isn't scheduled, protected, and expected, it disappears the moment pressure increases. Embedding it into the cadence made capability a

constant input instead of a future intention.

4. Energy management — in Navy, that often meant PT. In business, it looked like protected start times, finish times, or shared meals. The format changed with context, but the intent didn't. Sustained execution depends on managing energy deliberately, not assuming people will just cope indefinitely. If energy wasn't accounted for in the rhythm, output eventually paid the price.

5. Organisational obligations — mattered most for middle managers and anyone with reporting lines above them. Regulatory requirements, governance meetings, briefs, reporting cycles and standing commitments were known demands. When they weren't deliberately placed into the rhythm, they became constant interruptions. When they were planned for, preparation improved, energy was protected, and those obligations stopped hijacking the rest of the week.

6. Shared moments — marked progress. Sometimes that was recognition. Sometimes it was closing loops. Sometimes it was simply acknowledging that something difficult had been done well. These weren't rewards. They were signals. They reinforced what mattered and prevented effort from disappearing straight into the next demand.

This was never about running a perfect schedule. It was about coverage. When one of these components dropped out, pressure started getting absorbed through urgency and individual effort instead of structure. That was always the signal that the rhythm needed to be reset.

Keeping these six elements present ensured the team rhythm could carry tempo without constant intervention. Execution stayed organised because the system was doing its job, not because people were pushing harder.

A Real Hospitality and Catering Battle Rhythm (Why It Worked)

When I ran H&C, the environment was operationally heavy and unpredictable. Work didn't arrive neatly and problems didn't wait their turn. That meant the rhythm couldn't be clever, fragile, or overly detailed. It had to be simple enough to repeat and strong enough to hold under pressure.

I anchored the week around a small number of fixed points.

Monday was the reset. We ran a management alignment first thing, and that set intent for the entire week. Priorities were clarified, pressure points were surfaced early, and anything that mattered had a place to land. Nothing important was allowed to float.

I scheduled all one-on-ones on Mondays as well. That wasn't accidental. Those conversations gave me individual signal before the week gathered momentum. By Tuesday, I knew where people were overloaded, where support was required, and where I could safely apply pressure. That single decision removed most of the guesswork that normally drags leaders into reaction mode mid-week.

Energy management was deliberately built into the rhythm. In the military, that usually shows up as PT. In business environments the format can change, but the function doesn't. Movement, physical training, or shared activity stabilises energy over time. For my department, structured physical activity happened twice a week. It wasn't a perk. It was an input, and it paid for itself in resilience and output.

Training was another fixed point. One protected training window each week, non-negotiable. It wasn't reactive, and it wasn't delegated away. I ran most of the sessions myself because it mattered that development was visibly led, not outsourced. Sometimes, the focus was future capability. Other times, it was tightening execution around a process that was slipping. About once a month, a team member ran the session to reinforce learning and build depth through teaching.

Organisational obligations had defined places in the rhythm as well. Regulatory meetings, briefs, presentations, reporting cycles. These weren't surprises, so I treated them like known terrain. Placing them deliberately into weekly, fortnightly, or monthly cycles protected preparation time and stopped them from constantly interrupting operational work. Nothing undermines credibility faster than reshuffling meetings because you failed to plan for obligations you already knew existed.

Consistency in timing mattered more than flexibility. Once the rhythm settled, the team stopped guessing. People knew when issues would be heard, when changes would be discussed, and when I was focused versus unavailable. That clarity reduced noise immediately. Fewer interruptions. Fewer reactive conversations. Fewer escalations driven purely by uncertainty.

I also became deliberate with my own energy and calendar. I operate best early. Between seven and twelve-thirty I'm sharp. After lunch, I'm not. Once I accepted that, I scheduled heavy decision-making and complex work in the morning and pushed low-impact tasks into the afternoon. The quality of my output improved without adding hours or effort.

None of this was about control or rigidity. It was about removing uncertainty. Once the rhythm was clear, execution became calmer. Momentum stopped depending on constant intervention. The system carried more of the load.

That's why it worked.

Non-Negotiable Doesn't Mean Inflexible

This is where battle rhythm is most often misunderstood, and where people abandon it too early.

Non-negotiable doesn't mean rigid. It means protected.

The moment you cancel rhythm every time something urgent appears, you don't have a rhythm. You have good intentions that only survive calm weeks. As soon as pressure rises, the structure disappears, and the organisation slides straight back into reaction mode.

The discipline isn't in blindly following the calendar no matter what's happening. The discipline is in planning around the rhythm instead of constantly overriding it. When you do that consistently, your team adjusts their behaviour around it. When you don't, they won't, regardless of how often you talk about focus or priorities.

I've seen capable operators fail here more than anywhere else. Not because they lacked intelligence or effort, but because their time had no structure strong enough to resist interruption. Over time, that erosion showed up as fragmented attention, delayed decisions, and leadership that depended too heavily on availability instead of direction.

The OIC was a clear example. Her calendar was always full and publicly visible, but it reflected activity without structure. Meetings happened opportunistically wherever gaps appeared. Conversations were squeezed in without regard for timing or energy. There was no dedicated cadence for decision-making or preparation, and when senior leaders came looking for her and she wasn't available, the cost of that chaos landed on everyone else.

The issue wasn't effort. It was the absence of rhythm.

Different work draws on different energy. Meetings require one gear. Preparation, judgement, and deep work require another. When meetings are scattered across the week and focus blocks aren't protected, neither gets done properly. Attention fragments, availability becomes unpredictable, and leadership starts to feel reactive even when intent is sound.

This is exactly what battle rhythm is designed to prevent.

A full calendar doesn't mean an effective one. Visibility doesn't equal clarity. Without deliberate structure, time gets consumed by motion rather than progress, and leaders compensate with availability instead of intent.

That's why discipline in structuring time matters. Not to control people, but to protect your ability to be present where it counts. When commitments are clear and the rhythm is protected, decisions get easier. You stop saying yes out of urgency or guilt and start allocating time based on priority.

Jack Delosa has a line that has stuck with me:

Show me your calendar and your bank account and I'll show you what you value.

When I mapped every team, departmental, and organisational commitment into the calendar, what was left wasn't theoretical. It was confronting. There was far less discretionary time than I thought, which forced real trade-offs.

From there, the rules became simple. If something could be solved with a clear email, it didn't get a meeting. Meetings are expensive. They burn time, energy, and focus. And when a meeting was genuinely required, every decision-maker had to be present. Otherwise, it didn't happen.

This is where non-negotiable rhythm earns its value. It forces prioritisation, exposes waste, and makes trade-offs visible instead of hidden. Flexibility still exists inside that structure, but it's intentional. You bend the rhythm when it serves the mission, and don't break it every time something feels urgent.

Battle rhythm doesn't eliminate chaos. It stops chaos from deciding how your week runs. And once that happens, execution becomes sustainable instead of exhausting.

Adapting Rhythm to Different Environments

Battle rhythm isn't a template you copy from someone else's calendar. The shape of it has to match the environment you're leading. When it doesn't, rhythm turns performative. It looks organised but adds friction instead of removing it.

In operations-heavy teams, the tempo is high and the margin for drift is small. Work changes quickly, priorities shift, and small misalignments compound fast. These teams need frequent alignment and short cycles. The rhythm has to create regular reset points so issues surface early and decisions don't lag behind reality. Long, infrequent reviews don't work here. By the time you meet, you're already behind the work.

Project-based teams are different. Execution is driven by milestones, dependencies, and handovers rather than daily operations. The rhythm needs to anchor around delivery points and decision gates, not just the calendar week. Alignment matters most at transitions. Too much cadence creates noise. Too little, and risks only surface when they're expensive to fix.

Small teams need the simplest rhythm of all. They don't need ceremony, layers of meetings, or elaborate structure. They need clarity. A small number of predictable touchpoints where priorities are set, work is reviewed, and decisions are made quickly. Overstructuring a small team kills momentum faster than understructuring it, because the overhead starts to outweigh the value.

The mistake I see repeatedly is leaders copying someone else's rhythm without understanding why it works for them. What functions well for a founder, a large department, or a command team often fails completely when it's transplanted into a different context. The environment changes, but the rhythm doesn't, and friction follows.

The question is not "what meetings should I run?" The better questions is "where does execution tend to break down here?" Where do decisions need to land? Where does alignment drift? Where do problems show up too late?

When you design rhythm around those answers, the structure starts working for you instead of against you. The meetings that matter earn their place. The ones that don't fall away naturally. And the team stops relying on constant clarification because the system is doing more of the coordination.

That's the point of adapting rhythm properly. Not to look organised, but to make execution feel cleaner in the environment you're leading, not the one someone else is talking about.

The Point of This Layer, and Where It Takes Us Next

Battle rhythm isn't about adding work. It's about making the effort that already exists move in the same direction.

Without rhythm, even capable teams spend their time reacting to whatever lands next. Energy gets burned responding instead of progressing, and execution depends on how switched on the leader happens to be that week. With rhythm in place, effort stops scattering. Work moves forward because there's a predictable way it gets processed, reviewed, and adjusted.

This layer exists to take pressure off the individual leader. When execution relies on stamina, availability, or staying late to keep things together, performance is fragile. The moment attention shifts, things slide. Battle rhythm shifts that load into structure. Decisions have a place to land. Priorities get revisited before they drift. Momentum comes from cadence, not force.

This is also where execution changes texture. Issues don't get chased as they appear. They get handled in the right forum at the right time. Work stops being glued together through presence and starts being coordinated through rhythm. You're still accountable, but you're no longer the system.

Once rhythm is established, it becomes possible to manage work without suffocating it. You can see what's moving, what's stalling, and where adjustment is needed without being everywhere at once. The organisation doesn't need constant intervention to stay upright.

That's as far as rhythm can take you on its own.

The next layer is about daily systems. This is where execution actually runs day-to-day, supported by the cadence you've already built. It's where individual effort, team rhythm, and organisational intent meet in real time.

At that point, leadership stops being about holding things together and starts being about keeping them clean.

OPERATOR SYSTEM

THE MASTERY
Lead like a pro

THE EXECUTION
Build the rhythm

Level 9
Daily Systems = Self Management

Level 8
Battle Rhythms = Management

Level 7
One-On-One Meetings = Management

THE STRUCTURE
Build the how

Level 6
Individual Goal Setting = Leadership

Level 5
Department Goals = Management

Level 4
Accountability Charts = Leadership

THE FOUNDATION
Build the why

Level 3
Processes & Procedures = Management

Level 2
Standards & Expectations = Leadership

Level 1
Vision, Mission & Values = Leadership

Twenty-Four

Daily Systems — Managing Time and Priorities

Time management is really a misnomer; the challenge is not to manage time, but to manage ourselves.   — Stephen R. Covey

Have you ever felt like you weren't short on time, just behind?

Behind on work you know matters. Behind on decisions you've been avoiding because you don't have the space to think them through. Behind on recovery, reflection, and anything that doesn't scream loud enough to interrupt your day. The hours fill up, the weeks disappear, and the important work keeps getting pushed into tomorrow.

That was me.

When I wasn't managing myself, I had no right trying to manage anyone else. I didn't see it that way at the time. I told myself I was committed. Available. Carrying the load. What I was actually doing was reacting all day and calling it leadership. Being busy became the proof. Responsiveness became the metric. And because I was exhausted, I assumed I was doing something right.

The chaos around the OIC wasn't just about workload. It came from a lack of personal structure. I recognised it immediately because I'd already been living a version of it myself. That's why daily systems became part of how I operated from 2021 onward, not as a fix later, but as a line I refused to cross.

Battle rhythm fixed my weeks. It gave the organisation shape and stopped work from floating. Once that cadence was in place, something uncomfortable became obvious. I could see exactly how much time I had left to work with. Inside Navy, it was roughly twenty to twenty-three hours a week of genuine, controllable work.

That was the constraint.

At the same time, I was building and scaling Neptune. I wasn't choosing between roles. I was carrying both. The rhythm exposed the reality, but it didn't solve the problem. It didn't tell me how to use those hours properly once I had them. It just removed the excuses.

For a long time, I wasted them.

Before I built daily systems, I was reactive from the moment I woke up. Forty hours in Navy and what felt like another lifetime at Neptune. Always responding. Always interrupted. Always busy. Still behind.

One Tuesday morning still stands out. Just after 0500, phone in hand, replying to emails that had been sitting unanswered for days. A Neptune tender due in three days, untouched. Reporting season in full swing. Around fifteen reports hanging over me. Everything felt urgent. Everything felt fucking hard.

I was working nonstop.

And I wasn't moving anything that mattered.

That was the moment I separated motion from progress.

Battle rhythm decides when work gets prioritised at an organisational and team level. Daily systems decide who controls the day once that work exists. By the time I hit 2025, that distinction wasn't theoretical anymore. It was lived.

Under Brigg, I thrived because discipline and structure were enforced around me. When leadership changed and chaos returned, I tried to keep operating the same way. I held it together for a while. Not perfectly, but enough. Eventually, the environment broke me anyway. At H&C, when I rebuilt my own daily systems properly, I took back control.

Without daily systems, rhythm gives the team direction but leaves you in the passenger seat. The week looks organised, but your day gets hijacked. You're responsive instead of deliberate. Present instead of effective.

This chapter is about discipline — not motivational discipline, but operational discipline. Structure that decides where attention goes before the day gets a vote. Not to cope better, not to feel less stressed, but to take control of the work instead of letting it drag you around by the neck.

Once rhythm is in place, this is the layer where you stop surviving the day and start driving it.

What Daily Systems Do

Daily systems weren't about getting more out of myself. They were about protecting my attention.

When I didn't protect it, I stayed busy, reactive, and permanently behind, no matter how many hours I worked. When I finally did, output stopped being something I had to force. It followed naturally, because my attention was finally going where it mattered.

What worked for me was simple. A daily system that answered the same three questions every day, without negotiation.

The first question was: **what matters today?**

For a long time, I started my days scattered. I'd wake up already reacting. Checking emails. Thinking about what I'd missed. Carrying yesterday straight into today without deciding anything at all. By the time I arrived at work, I was already behind.

That changed when I started the day with myself.

That idea makes people uncomfortable. It used to make me uncomfortable too. It sounded selfish. What I saw was the opposite. When I didn't start the day intentionally, everyone else paid for it. I arrived rushed, underprepared, missing details, already reacting before the day had even begun. When I did start deliberately, I showed up clearer, calmer, and more useful to the people I was responsible for.

It's also the only part of the day you can control. Once the day gets moving, control disappears fast. I stopped giving away that window.

That's why Admiral William McRaven's *Make Your Bed* speech stuck with me.

Not because of the bed.

The point was simple. If you make your bed first thing in the morning, you've already completed something. You've already set a standard. And if the day turns to shit, you still come home to something done properly. It's not about discipline theatre. It's about setting a tone before the world gets a vote.

That idea became non-negotiable for me.

My own system reflects that. I start at 0445. These days that looks like getting up, weighing in, showering, making my pre-gym shake, and driving to the gym. I train until 0700, head back, go for a thirty-minute walk while listening to an audiobook, then home to shower, eat, and head into Neptune. I finish work around 1400 to 1500. From 1600 to 1800 I've got a running window depending on my training. Anything left becomes walking time or practical errands. Dinner is locked at 1830. From 1900 to 2000 is intentional learning. From 2000 to 2100 is proper switch-off. At 2100, I'm in bed.

People tell me it sounds boring. They ask where the fun is.

The answer is simple. It's scheduled.

Weekends are blocked for golf, mates, and reset. Training still happens because that's part of who I am. The freedom exists precisely because the structure does.

The second question daily systems answered was: **what can wait without causing damage?**

This is where a real priority framework mattered. I've used variations of the Eisenhower Matrix for years, but it only worked once I adapted it to reality. Not everything urgent is important, and not everything important needs attention today. Without a framework, the day gets hijacked by whatever feels loudest. With a framework, decisions stop being emotional and start being deliberate.

This is how I stopped stealing time from tomorrow just to survive today. I decided, consciously, what moved now and what could wait without creating a mess later. That shift alone changed how much pressure I carried through the week.

The third question was: **what should never have reached me in the first place?**

This is where management stopped being emotional and became mechanical in the best possible way.

If something landed with me that didn't need my judgement, authority, or accountability, it didn't get absorbed. It got pushed back using the **1–3–1 rule**. One problem. Three possible solutions. One recommendation. That rule forced thinking back down to where it belonged and stopped me becoming the dumping ground for unresolved work.

Rules replaced guilt. Defaults replaced debate.

Those rules weren't about controlling people. They were about protecting my attention and energy. Once they were in place, fewer things arrived at the wrong level, and when they did, they were easier to deal with.

Daily systems didn't rely on motivation or willpower. That kind of discipline runs out. They relied on defaults. When the defaults were set properly, I didn't have to fight myself every day. The system carried me forward even when energy dipped or the week got heavy.

James Clear says:

You don't rise to the level of your goals, you fall to the level of your systems.

That line is accurate because it's operationally true. When pressure hits, you don't magically become disciplined. You revert to whatever structure you've built.

That's what daily systems did for me.

They didn't make me exceptional. They stopped me bleeding attention on things that never deserved it in the first place. They created a floor I couldn't fall through.

In the Operator System, this layer is critical because it protects the leader. It protects your attention, your energy, and your ability to think. Without it, you're exposed. With it, you're in control.

And that's what makes everything else work.

Why This Layer Sits Where It Does in the Operator System

Battle rhythm allocates time at an organisational and team level. It decides when alignment happens, when decisions land, and where work gets processed collectively. Once the rhythm is set, the meetings are placed, and the obligations are accounted for, there is only a finite amount of time left that belongs to you.

That remaining time is where leadership either compounds or collapses. Daily systems exist to determine what happens inside it. Without them, the week can look disciplined while the day unravels. You spend what's left reacting, clearing noise, and absorbing work that should never have reached you. With daily systems in place, that leftover time becomes leverage instead of exposure.

This chapter sits here because this is the handoff point. Battle rhythm shapes the week. Daily systems decide whether you get to drive what remains. That's the difference between having a structured calendar and exercising operational control.

The Body Is Part of the System

In 2024, losing two organs forced a reckoning I'd spent years avoiding. I had built my identity around putting everything except myself first. Saying yes to everything. Carrying other people's pressure. Absorbing stress that was never mine to take on. I told myself it was leadership. It was people-pleasing dressed up as duty.

That way of operating nearly killed me.

In 2025, the choice was simple. Keep pretending I could outwork damage, or take control of my health before there was nothing left to recover. Once I got the all-clear from the medical team, I stopped negotiating with myself and started acting.

On 17 February 2025, I went for my first run.

Four kilometres. Nine minutes and twenty-nine seconds per kilometre. I'm well aware some people can walk faster than that. I didn't care. It was a starting point, and for the first time in a long time, the focus was on me.

By June, I ran the City to Surf in Darwin, twelve-and-a-half kilometres. In August, I ran the Sydney City to Surf at fourteen-and-a-half kilometres. In December, I ran twenty kilometres around Sydney. For some people, that's a warm-up. For me, it was real progress.

My left side is held together with history, hardware, and a bit of gaffer tape. I snapped my left Achilles tendon in November 2022 playing soccer for Navy FFA. I've had my left patella tendon replaced after it was effectively destroyed on deployment in 2018, when we made the brilliant decision to run the beep test every day on a hydrographic ship's flight deck. I got seriously fit. I paid for it later. That deployment with HS Red Crew was still one of the best sea postings of my career, even if my knees would disagree. My left shoulder has also been reconstructed after repeated dislocations, again from soccer.

None of this was about chasing an ideal version of fitness. It was about rebuilding a body that had taken years of punishment without care.

In May 2025, I started in the gym. My mate Kiarra coached me and handled my nutrition. The changes came faster than I expected because the foundations were finally right. In October, I stepped it up and signed on with Physique Factory. Kiarra built the base. Julian Baladron took me further.

Since coming out of hospital in December 2024, I've dropped over thirty-five kilos, lost more than twenty percent body fat, and rebuilt the capacity to run twenty kilometres. I no longer need heartburn medication. I understand calories for the first time in my life. But none of that is the most important change.

The biggest shift was mental clarity.

Dan Martell talks about it often:

Exhaust the body to tame the mind.

I understand that now in a way I never did before. If I don't do something physical, I feel off. My afternoon runs or walks reset me. They clear noise, settle my thinking, and let me come back to decisions with perspective instead of agitation. My morning sessions give me a platform for the entire day. They set the tone before the world gets involved.

This isn't about being fit. It's about being operational.

When the body is neglected, thinking degrades. When energy isn't managed, judgement suffers. When recovery is ignored, people compensate with adrenaline and stubbornness, and that pattern always ends the same way.

That's why this sits inside daily systems in the Operator System. Not as a luxury. Not as something you do if there's time left. It's structural. It underpins everything else. If you don't look after the machine you're running the system with, the system will fail, no matter how disciplined the calendar looks.

I don't train to look a certain way. I train so I can think clearly, lead properly, and carry responsibility without breaking.

That's why for me it's non-fucking-negotiable.

The Real Enemy: Other People's Priorities

This topic has been done to death, and I know it. Jack Delosa, Dan Martell, Taki Moore, Alex Hormozi. We've all talked about it publicly. At a CEO or founder level, the advice is simple: hand your inbox to an EA and move on.

Most managers don't get that option. If you're running a department day to day, deep in the work, you don't get separation. You get an inbox. And whether you like it or not, that inbox becomes the front door to your attention.

Your inbox is not your fucking job.

Emails, messages, forwarded tasks, CC'd conversations. None of them arrive labelled with what matters. They all ask for attention, and in the moment they feel urgent. Urgency starts deciding your day before you ever do. I've watched capable people spend entire days clearing messages and calling it productivity, only to finish exhausted with nothing meaningful moved.

That's the real enemy here. Not poor work ethic. Not lack of discipline. Other people's priorities, arriving unfiltered, all day, every day. If you've ever finished a week drained and wondered how you were so busy while still feeling behind, this is why.

The Priority Lens I Used

Most people know the urgent versus important idea. It's useful, but on its own it's incomplete. What changed everything for me was adding a filter that sat in front of it.

Significant.

For me, significant work was anything that moved the organisation, department, team, or my own stated goals forward. If a task didn't do that, it didn't go first, no matter how loud it felt when it arrived. That single distinction mattered because it forced me to decide what mattered before urgency ever got a vote.

Every email went through that lens first.

Is this significant? Does it move forward something I'm accountable for? If the answer was no, it was deprioritised. Not ignored, but stripped of emotional weight. It stopped feeling urgent simply because it existed.

Only after that did I apply the traditional categories.

Important and urgent items were handled immediately. These were genuine issues that couldn't wait without creating damage. Important but not urgent work was scheduled into the calendar. If it mattered but didn't need attention right now, it earned a specific time instead of floating in my head. Urgent but not important work was delegated. If I kept those, I was stealing time from work only I could do.

The final Eisenhower category is usually "delete". I didn't use it that way. Instead, I treated it as reference. Some messages weren't urgent or important but contained information I might need later. Those went into a small reference system rather than cluttering my inbox or my thinking.

I added one practical layer on top of that: anything that could genuinely be completed in under two minutes went into a separate folder. When I had a gap between tasks, I cleared it quickly without letting it bleed into deeper work.

The way I applied this was simple. When I logged in each day, I didn't respond or skim. I sorted. Emails were dragged into their respective folders until the inbox was empty. That process alone removed the sense of overwhelm because the inbox stopped being a pile of noise and became a set of decisions already made.

A side effect of this was that my inbox stayed empty most of the time. There's nothing revolutionary about that. Plenty of systems talk about inbox zero. I'm not selling a new idea here. I'm explaining what worked when I was managing departments and running a business at the same time.

Once the inbox was cleared, it stopped pulling at my attention. I wasn't scanning it constantly to see if something had arrived. I knew everything had already been filtered and parked where it belonged.

I worked from those folders during dedicated work blocks, not reactively throughout the day. Once items were completed, they were filed into a small number of functional folders. I capped myself at ten. That constraint was deliberate. Years earlier, I'd built a folder for everything, and finding information took longer than redoing the work. Fewer folders reduced friction immediately.

In Navy, those folders reflected reality: H&C Command, H&C Divisional, H&C Regulator, H&C Functions, H&C Safety, H&C Training, HMAS *Coonawarra* Command, Navy Football, and Personal Reference. At Neptune, the structure mirrored the business: Executive, Operations, Legal and Compliance, Finance, HR, Marketing, Business Development, Public Relations, Safety, and IT and Systems.

That structure worked because it matched how the work actually moved. It condensed my view, reduced mental load, and made retrieval automatic.

At the end of the day, I repeated the same sorting process. Anything new was filtered, parked, or scheduled. Doing this at close of business meant I never walked into a morning blind or reactive. The work was already shaped before the day started.

For me, the value was never the matrix itself. Plenty of people know the theory. The value was using it as a decision gate. Every message had to earn its place before it earned my attention. Once that rule was in place, other people's priorities stopped hijacking my day and my time started aligning with what I was actually responsible for.

That's what made the difference.

Daily To-Dos and Protecting Deep Work

Before I log on to any computer, I go old school. Pen and paper. No screens. No inbox. I write down what I need to achieve for the day, for myself and for the department. I run it through the same lens I use for email, but with one important constraint. At this point, I only care about what's significant and what's important. Everything else can wait.

A typical list might look simple. Preparation for H&C training. Check emails. Call Ozzie. Unfuck a supplier issue. That's it.

Out of that list, the most significant item is obvious. Preparation for H&C training moves the department forward, builds capability, and has downstream impact. That's the work that deserves protection. That's the work that goes into a deep work block.

This is where the system starts to come together.

I don't drift into deep work accidentally. I earn my way into it. Before any deep work block, I schedule a short window, usually around thirty minutes, to clear the inbox and knock over anything that genuinely takes two minutes or less. That stops small tasks from clawing at my attention once I'm trying to think properly. Once that window closes and I enter the deep work block, I'm unavailable. No interruptions. No quick questions. No just-this-once conversations. That boundary matters because without it, deep work never actually happens.

In my Navy role, I scheduled one deep work block a day. Most days, that was from 0900 to 1230. Mondays were the exception. Mondays were management meetings and one-on-ones, so the deep work block shifted to the afternoon, usually 1300 to 1530. The time moved, but the protection didn't.

There are a couple of ways to use these blocks properly. One is to dedicate them to significant work that directly advances stated goals. The other is to use them to work on the department or organisation itself, not just inside it. That distinction matters. At the entrepreneur level, people talk constantly about working on the business instead of in it because that's where leverage lives. The same principle applies in departments. If you never step out of the day-to-day, nothing improves. You just get better at coping.

For most middle managers, the reality is a mix. You still have a primary role to perform, and that fills much of the remaining time. The difference is that the significant work is no longer squeezed into leftovers. It gets a protected place. As you move higher in the management structure, those primary task blocks reduce and the leverage blocks increase. Lower down, it's the opposite. The system still works either way because it's designed around reality, not aspiration.

This is how I stopped confusing activity with progress. The day didn't decide what mattered. I did, before anything else got a chance to interfere.

In the next section, I'll walk you through what a full day actually looked like when this system was running properly, so you can see how all of these pieces work together in practice rather than theory.

A Full Monday, End to End

By the time I got serious about daily systems, one rule was non-negotiable: if it wasn't in the calendar, it didn't happen. That applied to work, training, recovery, travel, and even the margins most people pretend don't exist. The calendar wasn't a record of what I hoped to do. It was the operating system for how the day actually ran.

A Monday from my Navy calendar is a good example because it shows how everything fits together when the system is working.

The day started at 0445. Not because it sounded disciplined, but because it gave me uncontested time before the world started making demands. From wake-up through to pre-workout, travel, and the gym, that block was protected. Training ran from 0530 to 0700, followed by the return home, breakfast, uniform, and the commute to work. Even the coffee run was accounted for. None of that was accidental. When you pretend those transitions don't exist, the day starts late before you even realise it.

I arrived at work just before 0800, with a short window to talk shit with the lads. That mattered. Connection wasn't something I left to chance. Then I moved straight into intent. From 0800 to 0815, I wrote my daily to-do list. Pen and paper. No screens. No inbox. Just clarity around what actually mattered for the day.

From there, I gave myself fifteen minutes to sort emails. Not respond. Sort. Anything under two minutes got cleared. Everything else got filtered into the system I'd already built. That stopped the inbox from bleeding into the rest of the morning.

The next thirty minutes were spent preparing for the management meeting. Mondays were alignment-heavy by design. At 0900, the management meeting set direction for the week. Immediately after, I ran consecutive one-on-ones with my team and my own divisional officer. Those weren't scattered across the week. They were clustered intentionally. By 1030, I had a clear picture of where everyone was tracking and where pressure was building.

From 1030 to 1230, I ran what I'd call a short work block. This wasn't deep work. It was deliberate space for important and urgent tasks that didn't require heavy thinking. Problems surfaced in the morning meetings, and this block allowed me to deal with them before they became noise.

Lunch from 1230 to 1300 wasn't rushed or skipped. It mattered, and having H&C producing proper food made it even better. After lunch, there was a final fifteen-minute clearing window. Anything new that had arrived got filtered so it couldn't contaminate the afternoon.

From 1315 to 1530 was the deep work block. This was where significant work lived. No interruptions. No meetings. No quick questions. This was the time spent working on the department rather than inside it. Planning, preparation, system improvement, and thinking that actually moved outcomes forward.

At 1530, I shut it down and headed home. The workday ended cleanly. From 1600 to 1800, I had a running window. Some days that was a hard session. Other days it was a long run or a walk. The purpose was the same. Reset the system, clear the head, and burn off whatever the day had accumulated.

Even the evening followed a rhythm. Shower, dinner prep, dinner, and then a dedicated training and learning window. That learning wasn't passive. It was intentional. From there, a short window for posting or admin, then genuine downtime before bed at 2100.

Nothing about this day was exciting on its own. That was the point.

Because everything was accounted for, nothing had to be negotiated in the moment. I didn't waste energy deciding what to do next. I didn't feel pulled in ten directions. The system carried the day, which meant I could actually focus on execution instead of managing myself constantly.

This is what daily systems look like when they're working. Not rigid. Not impressive. Just deliberate. And once you experience a day like this running cleanly, it becomes very hard to go back to pretending chaos is normal.

Deep Work: The Fragile Part of Leadership

Deep work is fragile because it depends on discipline to survive. Not motivation or intent. The moment that discipline slips, the space starts shrinking. Meetings get moved into it. Interruptions creep in. Blocks get shortened or pushed. None of it feels catastrophic, but the boundary weakens.

Deep work doesn't lose because it's unimportant. It loses because it doesn't compete well with noise. Emails feel urgent. Meetings feel legitimate. Requests feel reasonable. Deep work doesn't announce itself, so if you don't protect it, it gets replaced. Once that replacement becomes normal, the quality of work drops even though activity stays high.

This is where leadership output quietly erodes. When deep work disappears, problems stop getting solved and start getting managed around. Systems get patched instead of improved. Decisions get made with less context because thinking time is squeezed into whatever space is left. Leaders stay busy, but leverage disappears.

Deep work only survives when it is treated as non-optional execution time. The moment it becomes movable, cancellable, or negotiable, it starts eroding. And as that erosion continues, so does the work that moves the organisation forward. That's why deep work is fragile. It doesn't fail because people don't understand its value. It fails because discipline stops being enforced.

Delegation Became Leverage, Not a Favour

Once my daily systems were in place, something else changed that I didn't expect. I stopped treating every decision like it deserved the same weight.

For a long time, I confused hesitation with responsibility. I'd sit on decisions, gather more input, revisit the same questions, and tell myself I was being thorough. What I was doing was slowing execution down and exporting uncertainty to everyone waiting on me.

Some decisions need care. Most need movement.

What fixed this wasn't confidence or instinct. It was structure. I started running decisions through a simple lens.

First, could the decision be reversed? If yes, it didn't deserve extended analysis. I made the call and moved. If it was hard to unwind, that's where I slowed down.

Second, what happens if I decide now? Not in theory, but operationally. What moves? Who gets clarity? What stops stalling?

Third, what happens if I wait? That question mattered most. Delays rarely preserve optionality. They create drag. Work stalls. People hedge. Small problems grow while everyone waits for permission to act.

Once I started deciding this way, speed stopped feeling reckless. It became deliberate. Fast decisions cleared small, reversible issues before they accumulated. Slower decisions were reserved for the few things that carried long-term consequence.

The effect wasn't subtle. Bottlenecks cleared. Hovering stopped. Momentum picked up because people weren't waiting on certainty that was never coming. Even when a decision wasn't perfect, movement created information, and information improved the next call.

This is where daily systems really earn their keep. They don't just protect time. They remove emotion from decision-making. Calls stop being shaped by mood, pressure, or who's asking loudest, and start being shaped by flow.

Perfection feels safe, but it's expensive. Velocity, applied properly, reduces load across the system. It keeps work moving and stops leadership from becoming the point where everything slows down.

That's when speed stopped being a risk and became a tool.

Interruptions: Defining What Matters

Interruptions never disappeared for me. What changed was which ones were allowed to cut through.

Before I had daily systems in place, everything felt interruption-worthy. People knocked on the door, sent messages, or grabbed me in the hallway because they didn't know where else to raise something. Urgency became subjective. Whoever arrived first got attention. That wasn't a leadership problem. It was a system problem.

Once expectations were clear, behaviour shifted without me needing to police it. The team learned what counted as a real interruption. Safety issues. Operational failures that couldn't wait. Decisions that genuinely blocked progress. Everything else had a place to land, whether that was a one-on-one, a scheduled forum, or a protected work block. When people knew there was a reliable place for issues to go, they stopped interrupting out of anxiety.

What surprised me was how many interruptions disappeared without confrontation. Not because I made myself unavailable, but because the system removed ambiguity. People didn't have to guess whether something was worth raising right now. The rhythm told them.

Interruptions still happened, but they were cleaner. When something cut through, it mattered. And because that was rare, it got full attention instead of irritation. That's how daily systems changed the environment. Not by fighting interruptions, but by defining what deserved to break the flow.

Technology Is an Amplifier, Not a Solution

I learned quickly that tools don't fix confusion. They expose it.

Calendars, task managers, CRMs, automation, AI. None of them worked for me until the underlying decisions were already made. Priority. Ownership. Intent. Without those, technology didn't help. It just made me faster at doing work that shouldn't have existed in the first place.

Early on, I tried to solve pressure with tools. New apps. New systems. New dashboards. The result wasn't clarity. It was noise. Tasks multiplied. Inputs increased. I felt productive while drifting further away from the outcomes I was responsible for.

Daily systems changed that. Once the rules were clear, technology finally had something useful to amplify. The calendar worked because priorities were already decided. Task lists worked because ownership was explicit. AI worked because I knew exactly what I was asking it to support, not because I expected it to think for me.

That distinction matters. Technology doesn't replace judgement. It accelerates whatever judgement you already have. If your thinking is sloppy, it will scale that sloppiness. If your intent is clear, it becomes leverage.

That's why tools come after systems in the Operator System. You don't start by choosing software. You start by deciding how work should move, what deserves attention, and what never should have reached you. Only then do tools earn their place.

For me, rules did the work. Not rigid rules, but decision rules. What gets done now. What gets scheduled. What gets delegated. What doesn't land with me at all. Once those were in place, technology stopped being distracting and started being useful.

Daily systems are thinking first, tools second. Reverse that order and you don't get leverage. You get speed without direction.

Used properly, technology supports the system. It never replaces it.

The Point of This Layer

This chapter isn't about productivity. I don't care how busy someone looks or how many tasks they tick off. It's about who controls attention.

Daily systems are what stop leadership energy being burned on noise and start directing it where it matters. Direction. People. Capability. Without them, battle rhythm exists on paper but falls apart in practice. The week has structure, but the days get hijacked. With daily systems in place, the cadence becomes usable. Structure turns into execution.

This is also the point where leadership stops requiring constant presence. You don't need to hover, chase updates, or stay late just to feel across things. The system does that work. Attention gets applied deliberately instead of reactively, and effort stops leaking into everything at once.

When daily systems are in place, the shift is simple but decisive. You stop trying to manage time and start deciding where energy and focus go. That's where self-management lives. Not motivation. Not willpower. These are operational standards that let you think clearly and carry responsibility without grinding yourself down.

This is Level 9 of the Operator System because this is where leadership effort starts to scale. The system continues to function when you're not checking, chasing, or holding it together by force. From here on, leadership stops being about coping with demand and starts being about directing it. Not control over people, but control over attention. That's the difference between surviving the workload and running the workload.

OPERATOR SYSTEM

THE MASTERY
Lead like a pro

THE EXECUTION
Build the rhythm

Level 9
Daily Systems = Self Management

Level 8
Battle Rhythms = Management

Level 7
One-On-One Meetings = Management

THE STRUCTURE
Build the how

Level 6
Individual Goal Setting = Leadership

Level 5
Department Goals = Management

Level 4
Accountability Charts = Leadership

THE FOUNDATION
Build the why

Level 3
Processes & Procedures = Management

Level 2
Standards & Expectations = Leadership

Level 1
Vision, Mission & Values = Leadership

Twenty-Five

Servant Leadership and Communication — Leading Through Service

Leadership is not about being in charge. It's about taking care of those in your charge. — Simon Sinek

I've never been particularly interested in most of what gets written about servant leadership. Not because it's wrong, but because it's usually framed in a way that doesn't match leadership under pressure. It gets presented as softness. Endless availability. Putting everyone else first at the expense of structure, clarity, and decision-making. That framing never matched how I led when things were moving fast and the margin for error was small.

What worked for me inside the Operator System wasn't a philosophy. It was a way of being useful.

Servant leadership, as I practised it, wasn't about being liked, selfless, or accommodating. It was about making execution cleaner for the people doing the work. It was about reducing friction, building capability, and stopping unnecessary weight from rolling downhill. Practical. Tactical. Built on top of a system that was already doing most of the heavy lifting.

The Operator System will carry execution regardless of leadership style. You could run it with a directive leader, a transactional leader, even someone bordering on a dictator, and on paper the system would still function. In the right conditions, someone like the OIC could probably look competent for longer than she did if the structure had been strong enough.

This chapter isn't about claiming servant leadership is the only way to lead. It's about how I chose to lead inside the system, and why it worked for me.

At the core, my role wasn't to be the centre of the system. It was to make the system work for the people inside it. That showed up in repeatable ways. I developed the people carrying the mission because execution only improved when capability did. I built capability by giving responsibility early, exposing people to real decisions, and giving feedback while it still mattered. I removed obstacles, whether that was bureaucracy, slow decisions, broken processes, or noise from above. And I pushed decision-making down instead of hoarding it, because if everything came through me, I wasn't leading. I was throttling the system.

None of that required softness. It required clarity, and clarity only holds if communication is disciplined.

This chapter sits here because the structure is already built. The rhythm is set. The daily systems are in place. This is how I led inside that structure, using service and communication to build clarity, capability, and momentum without becoming the point everything revolved around.

What Leadership Looked Like Under Pressure

After enough time under different leaders, patterns stop being theoretical. You don't need frameworks to tell you whether leadership is working. You see it in behaviour when pressure increases, when something breaks, and when no one is watching closely.

I saw it most clearly when I compare my two supervisors.

Under my first supervisor, the environment stayed coherent even when the work was demanding. Expectations were stable. Decisions didn't swing with mood or optics. Communication was clean and predictable. Pressure moved upward instead of rolling downhill. You knew what mattered and where you stood. Working for him wasn't easy, but it was logical.

My second supervisor's environment felt different. The focus wasn't on strengthening people or the system. It was on activity, presence, and appearance. Decision-making shifted. Communication was inconsistent. Structure was weak. Pressure accumulated at the lowest levels instead of being absorbed above. People spent more time managing the environment than executing within it, and capability didn't grow because ownership never stayed clear long enough to support it.

The difference wasn't rank, intelligence, or intent. It was how leadership showed up day after day. In the first instance, leadership reduced friction; in the second, it augmented friction.

That contrast reshaped how I understood servant leadership. It wasn't about being nice or endlessly available. It was about whether the way I led made it easier or harder for people to do their jobs. When leadership worked, it looked structured and demanding in the right places. People were trusted with responsibility and backed when it mattered. Decisions landed cleanly. Problems moved upward instead of spreading outward.

When leadership failed, it looked busy and reactive. People focused on reading the leader instead of doing the work. Energy went into managing uncertainty instead of executing outcomes. The leader became the bottleneck without ever meaning to.

That's the version of servant leadership that fit inside the Operator System for me. Useful. Disciplined. Built on clarity.

The Power of Positive Recognition

"Catch your staff doing something right at least 51% of the time!"

Early in my career, I was trained to look for what was wrong. Fix the problem. Correct the behaviour. Address the shortfall. Over time, that trains leaders to speak up only when something needs correcting.

The message people receive is simple. Silence means you're missing the standard. Visibility only comes with mistakes.

I saw what that produced. People didn't stop working hard. They stopped stretching. They stopped volunteering. They stopped caring at the edges, because there was no signal telling them what "good" looked like or whether it mattered.

That's where recognition changed for me. Recognition wasn't about being nice. It was a control mechanism. When people knew what good looked like and knew it would be noticed, behaviour stabilised without constant correction. Standards held without hovering.

When I started catching people doing things right in real time, the effect was immediate. Not exaggerated praise. Not public performance. Just specific acknowledgement. "I saw that." "That mattered." "That's the standard."

I aimed to recognise people doing things right more often than I corrected them doing things wrong. Not because standards didn't matter, but because recognition steered behaviour faster than correction ever did. It also changed how hard conversations landed. When recognition was normal, correction didn't feel like an attack. Accountability landed cleaner because the relationship could carry it.

Recognition costs nothing. It takes seconds. And the return is astronomical.

Tactical Communication Strategies

I didn't get better at communication by talking more. I got better by becoming deliberate about what I said, when I said it, and why I was saying it. People didn't need constant communication. They needed communication that held under pressure.

Learning to say no was the first shift. Most emails that land with a manager aren't information. They're other people's to-do lists. Early on, I accepted them by default and wondered why my priorities kept getting crowded out. Saying no wasn't about ego. It was about protecting what I was accountable for. Once people understood that, trust improved instead of degrading.

Learning to say "I don't know" mattered just as much. Pretending to know felt responsible at the time. It wasn't. Once I started being honest about uncertainty and committing to finding answers, credibility improved because the process was trustworthy.

When I did speak with certainty, I made sure it was conviction, not arrogance. Calm certainty carried more weight than volume ever did.

The most effective change I made was how I handled problems. Instead of solving them immediately, I asked: how would you solve this? At first it slowed things down. Then it changed everything. People started thinking before escalating. Judgement improved because it was expected.

That became the 1-3-1 rule. One problem. Three possible solutions. One recommendation. Not bureaucracy. Discipline. It forced clarity, reduced dumping, and made decisions faster.

And I ended most conversations the same way: what do you need from me? Sometimes the answer was a decision. Sometimes it was cover.

Sometimes it was access, permission, or reassurance. It kept my role focused on enabling action, not controlling it.

Feedback was the final piece. I gave it close to the event, positive or corrective, because delayed feedback loses relevance. When something went well, I named it early. When something needed correcting, I addressed it before it hardened.

These habits weren't impressive on their own. Applied consistently, they shaped the environment. Problems arrived with clarity. Decisions moved faster. Trust stabilised.

Development Through Opportunity

Most development doesn't fail because of money. It fails because leaders don't give people anything real to own.

The best development I ever saw cost nothing but time and attention. No courses. No consultants. No approvals. Just opportunity, backed by expectation. If someone wanted to grow, I gave them a place to prove it, and I made it clear I was backing them while they figured it out.

Reading and presenting was one tool. Someone would take a leadership or operational book, pull out what mattered, and brief the team on how it applied. Not a book report. Their judgement. Teaching forced deeper thinking and built confidence quickly.

Training delivery worked the same way. If someone cared about a topic, they built a session and delivered it. Standing up and teaching sharpened their thinking and exposed gaps safely.

The biggest lever was project ownership. When something needed fixing, I handed over the outcome. They scoped it, planned it, drove it, and briefed progress. I stayed available, but I didn't hover. Ownership builds judgement faster than formal training ever will.

I stretched people through deliberate scope expansion. One new area. One additional decision space. Enough to grow, not enough to crush them.

And if a process mattered, someone owned it intellectually. They became the reference point, kept it current, improved it, and trained others.

The key was that the opportunities had to be real. If you "delegate" responsibility but keep all decisions, people learn nothing. If you step in at the first sign of friction, you train dependence. Development only works when ownership is felt and support is consistent.

The Principles That Held Up

Over time, I stopped thinking about leadership as traits and started thinking about what still worked when resources were thin and scrutiny was high.

Vision mattered only if it showed up in behaviour. Integrity mattered in small, visible actions. Empathy mattered because pressure lands differently on different people. Accountability had to start with me. Adaptability mattered because plans change. Decisiveness mattered because waiting for perfect information creates drag. Communication mattered because clarity doesn't survive assumption. Empowerment mattered because leadership doesn't scale if everything bottlenecks around you. And learning mattered because leaders who think they're finished stop improving and the team pays for it.

None of that is impressive written down. The only thing that mattered was applying it consistently when it would have been easier not to.

Body Language, Presence, and Tone

I learned quickly that I was communicating before I said a word. People read presence in movement, posture, attention, and expression. If I rushed, fidgeted, or looked scattered, that energy spread even if my words were calm. If I moved with control and stayed present, the room steadied.

Eye contact mattered because it signalled attention. Facial expression mattered because it signalled safety. Posture mattered because it signalled whether the conversation was important. Small cues shaped what people brought to the table and what they kept to themselves.

Tone carried meaning before content ever landed. The same words could calm or escalate depending on how they were delivered. Variation kept people oriented. Pace signalled whether I was avoiding something or taking it seriously. Projection signalled whether I believed what I was saying. Lowering tone at the right moment slowed the room and added weight without aggression.

None of this was performance. It was part of the job. I couldn't demand calm while broadcasting chaos.

Building Your Leadership Legacy

I didn't think about legacy for a long time because it felt abstract. Eventually I realised legacy shows up in what your people do when you're not in the room.

Could someone on my team explain the direction accurately without me there? Would work keep moving if I wasn't available tomorrow, or would everything stall waiting for me? Did the team have judgement, or did they just have access to me?

I also had to get honest about my own gaps. The skill I wanted to avoid improving was usually the one costing my people the most. Feedback mattered, especially the kind that stung before it helped. Hard conversations mattered because avoidance never made things better. It just made them heavier.

If I stepped away tomorrow, what would be left behind? Not the projects. The people. Their confidence. Their judgement. Their willingness to take responsibility. Whether the environment I created made them stronger or smaller over time.

That's what servant leadership meant for me. Leaving people better equipped than I found them, and making sure clarity, trust, and capability survived my absence.

This chapter exists because servant leadership is the human layer that makes the Operator System sustainable. Without it, structure becomes rigid and people start compensating with effort. With it, the system doesn't just run. It lasts.

Twenty-Six

The Operator System Inside Hospitality and Catering

The ultimate measure of a man is not where he stands in moments of comfort and convenience, but where he stands at times of challenge and controversy. — Martin Luther King Jr

I've said it before, and I'll say it again.

In my honest, biased opinion, we built the highest-performing team in the Navy inside H&C in 2025.

That statement makes people uncomfortable because H&C isn't glamorous. It doesn't feature in recruiting videos or war stories told over beers. It's treated as background noise. Support work. Disposable. People reduce it to outputs: meals served, functions delivered, boxes ticked. They rarely look at what sits underneath it. Shift work that never stops. Short notice tasking. Chronic fatigue. Pressure that compounds daily and never gets applause because it's "just catering".

But performance has nothing to do with glamour. Performance is standards under load. It's whether a team executes when it's tired, understaffed, and getting pineappled by command week after week. It's how people behave when something breaks at the worst possible time, and whether they still deliver without needing to be dragged across the line.

That's what I mean by highest performing. A department that delivered day after day without drama. A team that could take a hit — and we took a fuckload of hits — adjust, and keep moving. A team that didn't need babysitting for the basics. A team you could trust. That's rare.

This isn't chest-beating or nostalgia. I didn't build that team through charisma or speeches. I didn't inspire anyone. We built it through standards, clarity, rhythm, and a culture that held when it was tested. People knew what mattered, what was tolerated, what wasn't, and what "good" looked like without it needing to be explained every week.

H&C is also where the Operator System stopped being an idea. It was inserted into a high-tempo environment where the workload didn't give a shit about staffing numbers. Command certainly didn't. Even when I got to ten chefs, the pressure didn't ease. The tasking kept coming. Function after function. Unrealistic expectations stacked on top of each other. And every time, the answer was the same. Smile. Endure. Deliver.

Most departments can hide behind complexity. H&C can't. Your output is immediate and unavoidable. People eat it. They feel it when it's late, wrong, or missing. Failure doesn't stay abstract. It lands straight on the plate.

That's why it mattered. If the Operator System couldn't work there, it didn't work at all. Not because H&C was broken, but because it was relentless. High tempo. Short staffed. Constantly under command pressure. If a system can hold in that environment, it isn't theoretical. It's proven.

Why Our Vision, Mission and Values Mattered

On the surface, what we did with Vision, Mission, and Values at H&C would look wanky to most Defence departments. Another poster. Another set of words rolled out on a divisional day and forgotten the moment real work starts. I understood the scepticism. I'd lived that version before.

I'd also seen what happens when departments rely on generic, top-down values that no one owns. The language sounds right, but it's broad enough to excuse almost any behaviour, including the ones quietly damaging the team. In those environments, culture becomes something people talk about instead of something that governs how work gets done.

That wasn't good enough for H&C.

The department had its own tempo, its own pressures, and its own failure points. Pretending a single, organisation-wide set of values could meaningfully guide behaviour there felt like avoiding the real work. Borrowing generic values is easy. Deciding who you are as a department isn't.

So we built our own identity around our organisation.

We defined what behaviour was expected when people were tired, busy, and under pressure. What was tolerated. What wasn't. What we protected when mistakes happened, and what we were prepared to confront when it would have been easier to look away. Once that was explicit, there was nowhere to hide. That applied to me as much as anyone else.

The difference was never the words themselves. It was how they were used. The values showed up in decisions, not slogans. They were referenced in hard conversations, not just praise. They explained why something wasn't acceptable, not only why something went well. Over time, people stopped asking what the values were because they could see them operating day to day.

At H&C, the values weren't aspirational. They weren't about who we hoped to be one day. They were behavioural. They described how we expected people to act when things were hard, because that's when values matter. Anyone can look aligned when conditions are good. Pressure exposes what really holds.

Most importantly, the values gave the team a reference point that wasn't me. Standards didn't live in my head. They were shared, named, and enforced consistently. That meant behaviour held even when leadership attention was stretched thin.

That's what saved the culture when pressure arrived from above.

When narratives started forming outside the department, the values gave the team language. They gave context to decisions. They gave people something solid to stand on that wasn't opinion or emotion. The system didn't survive because I was present. It survived because the values had already taken root.

That mattered more than I understood at the time.

Asking to Be Tested

In September 2025, I did something most leaders avoid.

If I was going to claim the Operator System worked, I couldn't keep relying on my own judgement. I needed to prove the hypothesis properly. That meant inviting scrutiny.

Not the polite kind. Not a walkthrough or a feel-good presentation.

I asked for a full cultural survey and training day to be run by the Director of Navy Culture, using the exact same framework, tools, and methodology that had been applied at FLSE-D in 2024 under the OIC.

That detail mattered.

It removed any room for me or the Operator System to hide.

This wasn't tailored to H&C. It wasn't softened to suit the environment or designed to make us look good. It was the same process that had exposed serious cultural failure elsewhere. If our culture was fragile, performative, or dependent on me holding it together by force, this would surface it quickly.

The difference wasn't the process.

The difference was why it was happening.

At FLSE-D, the intervention was reactive. The damage had already been done, and the survey followed it. At H&C, there was no trigger event. No complaint. No crisis. No external pressure.

I asked for it.

On the application, I stated plainly that I believed H&C was the highest performing and most culturally aligned team in the Navy. I didn't hedge it. I didn't soften it. I didn't bury it in qualifiers.

That wasn't arrogance. It was accountability.

If I was wrong, I wanted it exposed. If I'd been lying to myself, I wanted proof. I wasn't interested in protecting a narrative or preserving my ego. I wanted evidence. I wanted to know whether the system held when someone else controlled the lens.

There's an unspoken rule in Defence that you don't invite attention unless you're forced to. Scrutiny is treated as risk. Most leaders operate just below the radar and tell themselves that silence equals health.

I'd learned the hard way that it doesn't.

Seventeen of our twenty-one sailors attended the day. The absence of four mattered, but it didn't invalidate what followed. Those four were holding the department together in the background, cooking the food and keeping the wheels turning. That, in itself, said something about how the team functioned.

What mattered more was the tone.

This wasn't a forced evolution. People didn't arrive guarded or cynical. There were no filler activities or box-ticking exercises. The day was practical. Team-based problem solving. Real interaction. Observed behaviour. New exercises that made people think.

We started the day the same way we started every H&C gathering. With our values.

That alone shifted the room. The facilitators hadn't seen that before. Right there, it was clear they weren't dealing with a standard department.

After lunch, the 360-degree survey opened.

The same survey FLSE-D hadn't scored above fifty per cent in a single category.

At that point, there was no story left to manage. No framing. No way to explain outcomes away as misunderstanding or context. Whatever came back would reflect what the team experienced day to day.

I wasn't thinking about validation or reputation. I was thinking about responsibility. If the system worked, it deserved to be defended. If it didn't, it deserved to be dismantled and rebuilt.

That's what asking to be tested really was. Commitment.

The Results

I wasn't prepared for how stark the contrast was.

Not because I expected perfection, but because the difference was immediate and undeniable. The results were presented in the room, with all of us there. No delay. No private reflection. No time to frame it or soften it. What came back landed exactly as it was.

Out of two hundred and four total scores, five fell below fifty per cent.

Five.

That number matters on its own and even more in context. Using the same survey, the same methodology, and the same facilitators, FLSE-D had nearly three hundred out of three hundred scores under fifty per cent. Not a single category cleared the line.

This wasn't marginal improvement.

It was a different reality.

I didn't treat the five low scores as failure. I treated them as signal. Good data doesn't flatter or accuse. It points.

Three of the five sat in conflict. That told me something straight away. Much of what we were doing was landing, but conflict wasn't being experienced cleanly by everyone. I didn't know who had scored what, which meant I couldn't personalise it or explain it away. I had to treat it as a system issue, not an individual one.

One score sat in fun. That one stung.

Sixteen of the seventeen sailors scored that value a ten. The value itself — Bring fun, find happiness in every moment — was something you could feel in the workspace. But the single low score mattered more than the sixteen tens. It meant someone in the team wasn't experiencing what the rest were. A strong average doesn't cancel out a genuine outlier.

The final low score was opportunity. Someone felt they hadn't been given enough.

There was no defensiveness and no justification. That score told me there was at least one experience inside the system that didn't match the story I was telling myself. That's leadership information, whether it's comfortable or not.

Those five scores weren't problems to be dismissed. They were pressure points. Places to look, tighten, and improve. That's what healthy systems do. They show you where the load is building before cracks form.

But the section that stopped me cold was the values.

Every single sailor scored eight or above.

No fence-sitting. No quiet scepticism. The values weren't just recognised. They were trusted. People believed they were lived consistently enough to score them with confidence.

That was the moment I knew something had stuck.

Values are easy to write and easier to ignore. When they aren't reinforced, they fade. Seeing that level of alignment told me the values weren't symbolic. They were operational.

Then the facilitator spoke.

He said he sees requests like this all the time. Leaders claiming their teams are high-performing and culturally aligned. On paper, it looks convincing. In reality, it tends to unravel quickly once people start interacting.

He paused.

"Thank you for restoring my trust that cultures like this actually exist. What you've built here is rare, and you should be proud of it."

That mattered more than any score.

It didn't come from someone invested in protecting me, the department, or the narrative. It came from someone whose job was to cut through bullshit and see what was there. It confirmed that what we'd built didn't depend on my presence or personality. It was embedded deeply enough to withstand scrutiny.

That was the point where the Operator System stopped being something I believed in and became something I knew fucking worked.

The Bit People Don't Like to Talk About

H&C wasn't perfect.

Anyone who claims their team is flawless is either lying or not paying attention. There's a reason the old rule exists. Roughly ten percent of your people will consume ninety percent of your leadership time, energy, and emotional bandwidth. H&C was no different, and pretending otherwise would make everything else in this chapter meaningless.

I had that ten percent.

And if I'm being honest, I didn't handle every situation as well as I should have. Some conversations came later than they needed to. Some standards should have been enforced sooner. Some behaviours were tolerated longer than they deserved because I believed effort, care, and support would close the gap.

That part is on me.

One of the hardest lessons from this period was learning how quickly critical conversations lose power when you delay them, especially around performance. Avoiding those conversations isn't kindness. It's short-term comfort borrowed against long-term cost. You pay it back later through resentment, confusion, and drift.

You can do everything humanly possible for someone. You can support them, back them, train them, adjust workloads, and protect them from unnecessary pressure, and still end up as the villain in their version of events. That doesn't mean you failed. It means leadership does not come with narrative control.

That was difficult to accept.

Some people will never buy into values, no matter how clearly they are lived. Not because the values are wrong, but because they demand something the individual isn't willing to give. Values require change, adaptation, and accountability. Not everyone wants that trade.

In civilian organisations, that mismatch often resolves itself. People move on. Teams reshape. Fit matters.

In organisations like the military, it's rarely that clean.

Posting cycles, rank structures, and administrative reality mean you can be required to carry people who are fundamentally misaligned with the culture you are trying to build. You cannot always remove the problem. Sometimes all you can do is contain it, manage the impact, and protect the rest of the team from being dragged into it.

That creates tension.

You are building something coherent while knowing not everyone wants to be part of it. You are holding standards while being accused of being unfair. You are offering support while being told it isn't enough. That pressure never shows up in leadership case studies, but it's where much of the real work happens.

It's easy to celebrate success when things are going well. It's far harder to hold your nerve when the same system that delivered strong outcomes is tested by individual failure, resistance, or crisis.

That's where leadership stops being theoretical.

And that's where the next test arrived.

The Moment That Shook Me

In August 2025, just weeks before the culture survey, I received news no leader is ever prepared for.

One of my sailors had been admitted to hospital following a serious mental health crisis.

That sentence stays. It follows you long after the initial briefing, long after the procedural details are covered, long after you're expected to return to routine decision-making. Experience doesn't blunt it. Rank doesn't create distance.

This wasn't an abstract case or a line on a report. This was someone I had served alongside for years. Someone I knew well enough to recognise when they were no longer themselves. Their internal struggle hadn't appeared suddenly. It had been unfolding for nearly three years.

Earlier that year, I had raised concerns through the appropriate channels about whether I was the right person, or whether our environment was the right place, to support their recovery. I didn't believe I had the expertise required, and I was conscious of the limits of what leadership alone can provide in situations involving significant personal and medical complexity. I was directed to continue managing the situation within existing frameworks.

So I did.

The approach we took prioritised stability over output. Responsibilities were adjusted. Expectations were modified. Decisions were made to reduce exposure to additional pressure, both for him and for the team. These were not punitive decisions. They were attempts to balance duty of care, operational requirements, and the reality that I was operating within boundaries I did not control.

Like most complex situations, none of the contributing pressures existed in isolation. Work was only one element of a much broader personal context, much of which sat outside the department's visibility or influence. That complexity mattered, even when the system struggled to hold it.

After his hospitalisation, I attempted to make contact and was advised to step back and allow formal processes to take over. I complied. The situation moved beyond my direct involvement, and eventually the sailor was posted out of the unit.

For a time, I believed that distance might help. That the system would now provide the structure and support required.

What followed instead was scrutiny.

In November 2025, a formal fact-finding process was initiated into the department, focused on cultural concerns. This occurred weeks after independent survey data had reflected strong alignment and cohesion across the team. I recognised the pattern. When outcomes are confronting and causes are unclear, attention often shifts away from systems and toward individuals.

I engaged fully with the process.

At one point, I was asked to respond to an allegation about language I had used months earlier. I disputed the characterisation and explained the context in which the conversation occurred. The intent had been to set clear, realistic expectations tied to presence and responsibility, informed by feedback and documented plans. It was not delivered with hostility, dismissal, or malice.

Shortly after, I was informed of another serious incident involving the same sailor.

There is no clean way to carry that.

You replay conversations you believed were measured. You question decisions you thought were reasonable. You examine not just what you said, but how it might have been heard. You ask whether clarity caused harm, whether silence would have been safer, whether leadership itself became part of the weight someone was already carrying.

That was the point where the question turned inward.

Not whether the system worked.

But whether I had crossed a line I never intended to approach.

Not because the facts had changed, but because the outcome was devastating.

That moment forced a reckoning most leadership literature avoids. You can act with care, structure, and intent, and still be present when something goes profoundly wrong. You can follow process, seek advice, and operate within your authority, and still be left holding responsibility without resolution.

That doesn't absolve you.

It doesn't condemn you.

But it does change you.

It forces you to accept that leadership doesn't come with clean edges or definitive answers. Sometimes you never know whether a different choice would have changed anything. Sometimes all that remains is the obligation to reflect honestly, to keep interrogating your own judgement, and to resist the comfort of simple explanations.

That was the moment that shook me.

Not because it disproved the system.

But because it showed me that you can do everything within your control as a middle manager, act with care and intent, and still be left carrying consequences you will never fully resolve.

Where I Landed

I didn't replay this period endlessly or lose sleep questioning every decision.

I was confident in the approach I took and the way I handled it. When I stripped it back to the one question that mattered — would I change anything — the answer was simple.

I would have pushed harder, earlier, to get the member out of the department.

Not out of frustration. Not out of blame. But out of responsibility. Keeping someone in an environment that could no longer support them was not care. It was exposure. That was the line I should have drawn sooner.

That is the lesson I took from my final Navy posting.

Leadership and management are not skills you master and move on from. They shift because people shift. Context changes. Pressure reshapes everything. What works with one team, in one moment, under one set of conditions, can fail badly somewhere else if it's applied without judgement.

That's where leaders get into trouble.

Not through intent or malice, but through rigidity. Through assuming that past success guarantees future correctness. Through repeating approaches that once worked without noticing that the environment, or the people, have changed. That's how damage gets done while leaders still believe they're acting in good faith.

Improving my leadership was never a destination. It was a discipline. It meant choosing reflection over defensiveness. Owning decisions without softening them. Holding standards even when they created short-term discomfort, because I believed they protected the team in the long run.

Sometimes that meant backing people hard.

Sometimes it meant making calls that didn't feel good.

Sometimes it meant accepting that certainty isn't always available.

One of the hardest lines to hold was between compassion and responsibility. Compassion without standards turns into avoidance. Responsibility without compassion turns into harm. That balance is never fixed. It shifts depending on the person, the moment, and the context. Getting it wrong doesn't make you malicious. Refusing to examine it does.

At H&C, the Operator System worked.

The outcomes were real. Performance held. Retention improved. Pride and cohesion were visible. Culture wasn't something we talked about. It was something the team lived. They owned it. They defended it. They were proud of it.

I was proud of them. That pride didn't remove the weight. It sat alongside it.

Leadership isn't measured only by the outcomes you can quantify. It's also measured by what you carry when things go wrong, even when you didn't cause them, even when you acted with care and intent using the information you had at the time.

That doesn't invalidate the work. It acknowledges the reality.

There are no clean answers. No framework removes the human risk of leading other human beings. Avoiding that responsibility doesn't make you safer. It just makes you absent.

The Operator System didn't make me flawless. It didn't protect me from pain or failure. What it did was give me a way to lead with intent instead of chaos. To build structure without stripping out humanity. To create environments where people had the best possible chance to perform, grow, and hold the line together.

Sometimes that is enough.

Sometimes it isn't.

Leadership isn't the highlight reels. It isn't the slogans or the reports that smooth out the edges. It's living with the consequences of decisions made under pressure, with incomplete information, and still showing up to do the work again.

The Last Test

My final day in the Navy wasn't meant to be dramatic.

Monday, 19 January 2026. A quiet finish. Team breakfast at Fresh Point Parap, our spot. Coffee. Laughter. The usual piss-take. Then a slow wander toward the Neptune office, four hundred metres away, letting my head drift because for the first time in twenty years, I didn't have to rush anywhere.

The system had other ideas.

Before the coffee had time to settle, the new Divisional Officer, Mitch Litch took a call. All H&C staff were to report immediately to the Naval Police Coxswains Office for urinalysis. No warning. No explanation. Just turn up and piss in a cup.

I wasn't worried. I hadn't partied early. I don't do that shit. Still, there was something fitting about it. Twenty years of service, and the last administrative act was proving I wasn't high.

Negative result. Clean. Predictable.

I went home, logged on, and started tidying emails. Closing loops. Final handovers. Mentally already half out the door. Then another message landed.

Report back at 1330 to close out the fact find into H&C culture.

That's what I thought it was. It wasn't.

I walked into the room expecting a procedural close. Instead, I was informed that the fact find had been into me the entire time. At no point during the process had anyone bothered to mention that.

The explanation was almost casual.

"Well, you're the Chief. So, you're automatically the respondent." Right.

Five complaints of unacceptable behaviour.

Five.

One was raised straight away. I'd allegedly created a "boys' club."

This was said out loud. With a straight face. In a department where two of my most critical roles were held by women whom I trusted explicitly and backed relentlessly. The accusation collapsed under its own weight before it even finished leaving his mouth.

The Commander didn't drag it out. He said the word I'd spent most of 2024 hating.

Unsubstantiated.

That word had nearly broken me eighteen months earlier. That day, it landed differently. It felt like air returning to a room that had been sealed for too long.

Then he added something that mattered more than the finding itself. With all the facts considered, he said I had simply used my key people in key roles to get the outcomes required. That any leader placed in the same situation would have done exactly the same thing.

That was it. No warning. No spectacle. No redemption speech. Just a clean statement of reality.

On my final day in uniform, the system ran one last test. Not of culture, or process, or compliance, but of judgement.

And it passed.

Not because I was liked. Not because I played politics. But because the work held. The people held. The outcomes held.

I walked out of that building knowing something with absolute certainty. The Operator System didn't just survive scrutiny when I invited it. It survived scrutiny when I wasn't in control of the room.

That's the difference between a leader holding a culture together by force, and a system that actually works.

Epilogue

After the System

What I believe now is simpler, and heavier.

Leadership is not about how much pressure you can absorb. It's about how much pressure you prevent from landing where it shouldn't. It's about building systems that catch weight early, before it turns into personal damage. And when you fail at that, as you inevitably will, it's about having the honesty to face the cost without pretending it was noble.

I still carry people with me. Not sentimentally, but practically. I carry the sailors who trusted me. The ones who thrived and the ones who struggled. The ones who stayed and the ones who didn't.

I carry conversations I wish I'd had earlier. I carry decisions that were operationally right and emotionally devastating. I carry moments where I don't know whether a different sentence, a different tone, or a different moment would have changed anything at all. That ambiguity doesn't resolve itself with time. You don't get over it. You learn to live alongside it without letting it harden into either guilt or arrogance.

That's a part of the job no one really prepares you for.

The Operator System didn't remove that weight. It never could. What it did was stop me from pretending the weight didn't exist. It gave me a way to design environments where fewer people had to carry it alone. It gave me structure when chaos was the default, clarity when noise was loudest, and language that didn't rely on slogans, rank, or reassurance. It also forced me to confront my own limits, something no framework will do for you unless you're willing to let it.

There's a temptation at the end of a book like this to look for redemption. To suggest the pain was necessary, that everything happened for a reason, that the outcomes justify the cost. I won't do that.

Some things shouldn't have happened. Some costs were too high. Some people paid prices they didn't deserve. Acknowledging that doesn't weaken the system. It strengthens it. Any leadership model that requires denial to survive is already lying to you.

If you take anything from this book, I hope it isn't admiration or agreement. I hope it's responsibility. Responsibility to be honest about the systems you're operating inside. Responsibility to notice when people are compensating for structural failure. Responsibility to stop confusing motion with progress, or control with care. Responsibility to build things that don't collapse when you're tired, distracted, or gone.

Because leadership is temporary. Every posting ends. Every role turns over. Every name on the door is replaced. What remains is what you built while you were there.

Did people know what mattered without you saying it? Did standards hold when you weren't watching? Did decisions still move, or did everything bottleneck around you? Did pressure travel upward, or did it crush the middle? Those answers are your legacy, whether anyone remembers your name or not.

This system isn't mine once you close this book. Use it. Change it. Break it apart and rebuild it for your context. Discard what doesn't fit. Keep what holds. Just don't pretend that leadership is lighter than it is, or that systems are optional when pressure is real.

You don't get to avoid the weight. You only get to decide whether you carry it alone, or build something that can carry it with you.

That's where I've landed. Not certain. Not finished. Just clearer about the responsibility that comes with asking other people to follow you.

And clearer about this. If anything here survives me, it won't be the stories. It will be the systems that kept working when I wasn't there to hold them together by force.

This is no longer mine. It's yours.

About the Author

Matthew Jacques served for twenty years in the Royal Australian Navy, reaching the rank of Chief Petty Officer and holding leadership roles in high-pressure operational environments.

In 2017, while still serving, he founded Neptune Holdings Group and built it alongside his naval career. The company grew inside real constraint, with limited authority, limited time, and competing responsibilities. It was not a post-service reinvention, but a business built in parallel with active service.

After leaving the Navy, Matthew continued to lead Neptune Holdings Group as CEO, applying the same operator logic required to function inside institutional systems that do not bend for individual leaders.

The Operator System emerged from lived experience across both worlds. It was shaped by running teams where outcomes mattered, pressure was constant, and failure carried consequence, often without the authority, protection, or support leadership theory assumes.

He lives in Australia.